Nanostructures in Electronics and Photonics

Nanostructures in Electronics and Photonics

edited by
FAIZ RAHMAN
University of Glasgow, UK

PAN STANFORD PUBLISHING

Published by

Pan Stanford Publishing Pte. Ltd.
5 Toh Tuck Link
Singapore 596224

Distributed by

World Scientific Publishing Co. Pte. Ltd.
5 Toh Tuck Link, Singapore 596224
USA office: 27 Warren Street, Suite 401-402, Hackensack, NJ 07601
UK office: 57 Shelton Street, Covent Garden, London WC2H 9HE

British Library Cataloguing-in-Publication Data
A catalogue record for this book is available from the British Library.

NANOSTRUCTURES IN ELECTRONICS AND PHOTONICS

ISBN-13 978-981-4241-10-6
ISBN-10 981-4241-10-5

Printed in Singapore by Mainland Press Pte Ltd

PREFACE

For many years now, engineering at the very small scale has remained the driving force behind a significant share of the global economy. Over the past four decades, microelectronic devices have proliferated and their myriad uses have increasingly defined our modern lifestyles. Throughout the microelectronics era, the sizes of individual devices integrated on semiconductor chips have been steadily reduced. With this trend still continuing, a future is in sight when we shall reach the end of the road for conventional device miniaturization. This will come about both as a result of the graininess of matter and the quantum mechanical nature of physical phenomena that become evident at very small length scales. Although the limitations would principally come from material science and physics, their repercussions would be largely economical. In order to keep the semiconductor industry profitable, new materials, processing techniques and device architectures will be needed. At the time of this writing we have about ten years to prepare for a graceful handover to alternative technologies. By all accounts, these futuristic technologies will exploit the physics and technology of novel devices whose dimensions will be measured in nanometers — 1 nanometer being one-billionth of a meter. Structures and devices at this scale are already at the heart of modern technology and these are going to play increasingly important roles in the future. Nanotechnology — the engineering discipline concerned with studying and fabricating such ultra small objects is now a thriving field in applied sciences and is attracting more researchers and funding throughout the world. Several countries have made nanotechnology funding a very high priority in their budget allocations, as an appreciation for its potential has grown in recent years. Developed countries as well as a number of developing countries see much promise from nanotechnology in fields as diverse as information technology and health care.

The interdisciplinary nature of nanotechnology is perhaps the most noticeable feature of this new field. The nano world can be approached from several different disciplines such as materials science, electronics, physics, chemistry and biology. This creates new opportunities for scientists and engineers as well as policy makers. It is clear that explorations and innovations in this realm will open up entirely new possibilities. This is good news for a world that is increasingly short of non-renewable sources of both materials and energy. Recognizing the potential of nanotechnology, many companies have started research in this field and more are joining them every year. Universities too are increasingly active in this area. As researchers we need to have a broad understanding of what our fellow researchers are doing elsewhere. This book is geared towards satisfying that need. Nanotechnology being such a wide discipline, this book is only concerned with its applications in electronics and photonics and in that too it looks at only a narrow selection of topics. The book contains a selection of 16 chapters contributed by a number of research teams around the world. They have especially expanded and adapted these chapters from papers published by them in the recent past.

After the introductory chapter the book is divided into two parts. The first is comprised of works that deal with electronic applications of nanotechnology whereas the following part is constituted of nanotechnology applications in pushing the frontiers of photonic technologies. The chapters are focused on experimental aspects of nanotechnology rather than theoretical studies or computer modeling. All the contributors are active researchers in their fields of specialization and thus this book provides an up-to-date survey of the state of contemporary nanotechnology. The publisher, editor and contributors hope that it will be useful to both students and professional researchers alike.

Editor
Faiz Rahman
Department of Electronics and Electrical Engineering
University of Glasgow
Scotland, United Kingdom

CONTENTS

Nanoscale Materials and Structures for Photonics

LIST OF CONTRIBUTIONS

1. From Microstructures to Nanostructures

Faiz Rahman

Department of Electronics and Electrical Engineering, University of Glasgow, Rankine Building, Oakfield Avenue, Glasgow G12 8LT, UK

NANOSCALE MATERIALS AND STRUCTURES FOR ELECTRONICS

2. Assembling Ferromagnetic Single-electron Transistors with Atomic Force Microscopy

Hakån Pettersson[a,b], Ruisheng S. Liu[a,b], Dmitry Suyatin[b] and Lars Samuelson[b]

[a] Center for Applied Mathematics and Physics, Halmstad University, Box 823, SE-301 18, Halmstad, Sweden

[b] Solid State Physics and the Nanometer Structure Consortium, Lund University, Box 118, SE-221 00 Lund, Sweden

3. Nanoporous Alumina Templates for Nanowire Electron Devices

Travis L. Wade[a], Costel S. Cojocaru[a], Jean-Eric Wegrowe[a] and Didier Pribat[b]

[a] Laboratoires des Solides Irradiés, Ecole Polytechnique, 91128 Palaiseau, France

[b] Laboratoire de Physique des Interfaces et des Couches Minces, Ecole Polytechnique, 91128 Palaiseau, France

4. Single-walled Carbon Nanotube Transistors

Sunkook Kim and Saeed Mohammadi

Birck Nanotechnology Center, Purdue University, West Lafayette, IN 47907, USA

5. Cooling with Integrated Carbon Nanotube Films

Géza Tóth[a], Krisztián Kordás[a], Pulickel M. Ajayan[b,c] and Robert Vajtai[d]

[a] Microelectronics and Materials Physics Laboratories, Department of Electrical and Information Engineering, and EMPART research group of Infotech Oulu, P.O. Box 4500, FIN-90014 University of Oulu, Finland

[b] Department of Materials Science & Engineering, Rensselaer Polytechnic Institute, Troy, New York 12180, USA

[c] Department of Mechanical Engineering and Materials Science, Rice University, Houston, Texas 77251-1892, USA

[d] Rensselaer Nanotechnology Center, Rensselaer Polytechnic Institute, Troy, New York 12180, USA

6. AC Dielectrophoresis Alignment of Gallium Nitride Nanowires (GaN NWs) for Use in Device Applications

Sang-Kwon Lee, Tae-Hong Kim and Seung-Yong Lee

Department of Semiconductor Science and Technology, Chonbuk National University, Jeonju, Korea

7. Design, Fabrication, and Applications of Large-area Well-ordered Dense-array Three-dimensional Nanostructures

Chang-Hwan Choi[a] and Chang-Jin "CJ" Kim[b]

[a] Mechanical Engineering Department, Stevens Institute of Technology, Hoboken, New Jersey, 07030, USA

[b] Mechanical and Aerospace Engineering Department, University of California at Los Angeles (UCLA), Los Angeles, California 90095, USA

8. UV-NIL Stamp Fabrication Techniques with Diamond-like Carbon Film

Jun-ho Jeong, Ali Ozhan Altun and Ki-don Kim

Nano-Mechanical Systems Research Center, Korea Institute of Machinery and Materials, 171 Jang-dong, Yuseong-gu, Daejeon 305-343, Korea

9. ZnO Nanowires and Nanobelts: Structure Switch by Indium Doping

Hong Jin Fan

Max Planck Institute of Microstructure Physics, Weinberg 2, 06120 Halle, Germany

10. Field Emission Properties of 1-D SiC Nanostructures

Guozen Shen[a] and Di Chen[b]

[a] Nanoscale Materials Center, National Institute for Materials Science, Namiki 1-1, Tsukuba, Ibaraki 305-0044, Japan

[b] Photocatalytic Materials Center, National Institute for Materials Science, Sengen 1-2-1, Tsukuba, Ibaraki 305-0047, Japan

NANOSCALE MATERIALS AND STRUCTURES FOR PHOTONICS

11. Manipulating the Optical Properties of Individual and Arrays of Gold Nanopyramids

Jeunghoon Lee, Joel Henzie and Teri W. Odom

Northwestern University, 2145 Sheridan Road, Evanston, IL 60208, USA

12. **Properties of Gold Nanoantennas in the Infrared**

F. Neubrech[a], J. Aizpurua[b], S. Karim[c], T. W. Cornelius[d], and A. Pucci[a]

[a] Kirchhoff-Institut für Physik, University of Heidelberg, Im Neuenheimer Feld 227, D-69120 Heidelberg, Germany

[b] Donostia International Physics Center, Donostia-San Sebastian 20018, Spain

[c] University of Marburg, Hans-Meerwein-Str., D-35032 Marburg, Germany

[d] Gesellschaft für Schwerionenforschung (GSI), Planckstr.1, D-64291 Darmstadt, Germany

13. Three-dimensional Holographic Polymeric Photonic Crystal Operating in the Optical Communication Window

Jiaqi Chen and Ray T. Chen

Microelectronics Research Center, Department of Electrical and Computer Engineering, The University of Texas at Austin, Austin, Texas 78758, USA

14. Continuous Roll Nano-imprinting Technology for Large-scale Nano- and Microstructures

Shinill Kang, Suho Ann and Seok-min Kim

School of Mechanical Engineering, Yonsei University, 134 Shinchon-dong, Seodaemoon-ku, Seoul 120-749, Korea

15. Fabrication and Characterization of Two-dimensional ZnO Photonic Nanostructures

Jingbiao Cui

Department of Physics and Astronomy, University of Arkansas at Little Rock, Little Rock, AR 72204, USA

16. Visible Light Emission from Innate Silicon Nanoparticles in Silicon-compound Films Grown at Low Temperatures

Cao Zexian

Beijing National Laboratory for Condensed Matter Physics, Institute of Physics, Chinese Academy of Sciences, PO Box 603, Beijing 100080, People's Republic of China

1

FROM MICROSTRUCTURES TO NANOSTRUCTURES

Faiz Rahman

This introductory chapter is a brief survey of the microfabrication technologies that power today's semiconductor industries. The economic and technological trends have been described with both a medium term and a long term outlook. We examine some of the most important process technologies deployed in fabs across the world and look at the likely ways that these technologies will change in the future. We also investigate contemporary research that is creating a base for commercial nanotechnologies of the future. Following chapters take a more detailed look at some selected techniques for the fabrication of nanostructures.

In the early years of the last century no one could have predicted that one day engineering will come to depend so heavily on artificial structures that are thousands of times smaller than the diameter of a human hair. Indeed, a hundred years ago, 'thin as hair' was a frequently used catch phrase. Engineering in those days was associated with large and often very large structures; both mechanical and architectural. Even vacuum tube-based electrical engineering didn't deal with particularly small devices. The development of the first Germanium bipolar transistor in 1947 started a new paradigm shift with the first inklings of miniaturisation in electrical engineering. However, it took another few years before the idea of printing devices on a piece of silicon took shape and ushered in the era of the integrated circuit that started the microelectronics revolution in earnest. From that time on, integrated semiconductor devices have been continually shrunk to ever smaller sizes in order to improve product functionality and return on manufacturing investments. There is still much steam left. Semiconductor industry will keep forging ahead in much the same way as it is doing today until at least 2018. Beyond that, radical new technologies will be needed to sustain the same pace of development. Some of this will, undoubtedly, be the technologies that are being researched today in academic and industrial R&D labs across the world. Exactly when the transition will take place is anybody's guess but that it will happen is certain. This is a good enough reason why we should be interested in contemporary developments in nanotechnology but there is another dimension to it.

Since the dawn of solid-state electronics in the late nineteen-forties there has been relentless miniaturisation of active devices. The development of integrated circuits in the early nineteen-sixties ushered in the micro era with the advent of terms such as micro-circuits and micro-electronics. Typically, these referred to individual devices having dimensions of a few tens of micrometers. Some of the techniques for making small structures in solid materials had existed for many years but a great deal of the fabrication technology was specifically developed to construct monolithic integrated devices measuring only a few microns across. Very rapidly, these techniques were improved upon to make still smaller devices thus packing more devices on a chip of given size. Gordon Moore, one of the founders of Intel Corporation predicted in 1965 that the number of transistors on a chip will keep doubling every eighteen to twenty-four months. This prediction, known as Moore's law, has remained strikingly valid ever since and has guided decades of efforts in monolithic integration. Nowadays, with mainstream advanced microprocessor and Digital Signal Processor (DSP) chips boasting of transistor counts in excess of one billion, the success of Moore's law in predicting the evolution of the IC industry is more evident than ever before. A plot of rising transistor count per chip over the years since 1970 is shown in Figure 1.1.

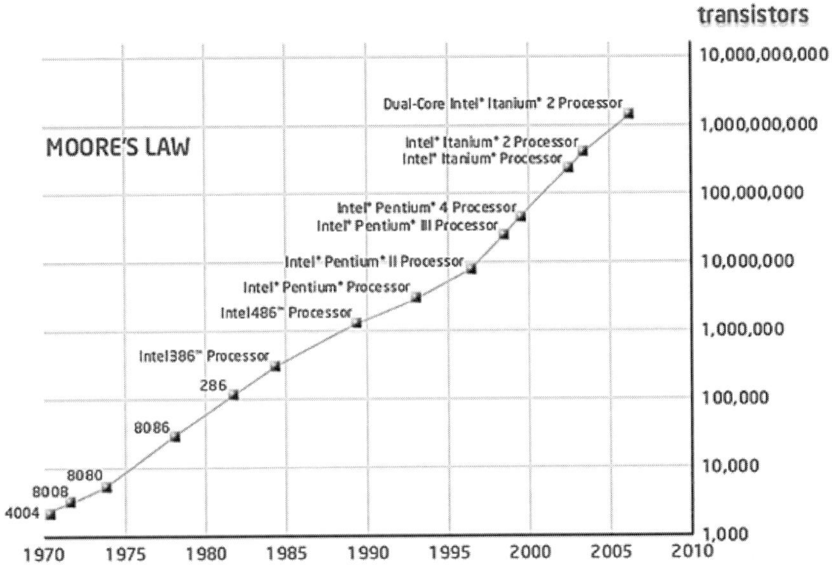

Figure 1.1 Transistor count for various Intel chips versus year of introduction (Courtesy: Intel Corporation).

During the sixties and the seventies device dimensions gradually shrank down to a few micrometers. By that time the benefits of miniaturisation were well understood and there developed a separate industry of miniaturisation tool suppliers. As more advanced technologies for next generation semiconductor chip fabrication were perfected, device miniaturisation marched on to the micron and then the sub-micron range. These were made possible by the development of such processes as dry etching and high resolution optical lithography. With device dimensions continually being pushed down deeper into the sub-micron range we entered the nano era around the turn of the present century. At the time of this writing, 65 nm devices are in commercial production and 45 nm devices are planned for release soon. The plan for three-year cycle technology generations or 'nodes' is shown in the table below. A node is defined as one-half of the smallest pitch in the technology. This information is taken from the International Technology Roadmap for Semiconductors (ITRS) which is a consensus reference document with a 15 year outlook on the requirements of the semiconductor industry (Table 1.1).

Table 1.1 ITRS silicon technology transition roadmap.

Production Year	2001	2004	2007	2010	2013	2016
Technology Node (nm)	130	90	65	45	32	22

There are sound business reasons driving this push towards smaller and smaller device geometries. As transistors shrink in size, a given number of them can be packed in a smaller area thus reducing the physical size of chips. This essentially means that a larger number of chips can be produced from a wafer, thus increasing the profits for the semiconductor companies. Of course, this holds true only if the chips per wafer yield is also maintained at the same time – no mean feat as features get ever smaller. Yet another way the shift to smaller device sizes has a favourable affect on a company's bottom line is through the possibility of packing more transistors in a chip of the same size as before. This way, manufacturers could pack more functionality in the same sized die as before, enabling them to fabricate and sell more capable integrated circuits that fetch higher prices. These were the factors that caused semiconductor companies to invest heavily in new fabrication tools and technologies in order to enable them to keep migrating from one IC generation to the next, every two to three years. And, as a consequence, the cost of individual transistors on monolithic chips have plunged many orders of magnitude over the past four decades – from around two dollars a transistor in 1968 to less than a nano dollar a transistor in 2008!

Many semiconductor foundries around the world are operating round the clock to produce an extremely wide range of analogue and digital chips. The 24 hour operation is necessary to extract the full return-on-investment from the installed equipment base. Figure 1.2 here is a view inside one of Intel Corporation's manufacturing facilities.

Another way the industry has been raising its profits is through the use of larger wafers. A few years ago six-inch diameter wafers were used for IC fabrication. Then the industry changed to eight-inch wafers and at the time of this writing twelve-inch diameter silicon wafers are the standard in leading fabs. These are planned to be displaced by eighteen-inch wafers by 2012 and there are talks of transitions to even larger wafer sizes in the future. Processing larger wafers is more expensive both because it requires new or modified tools capable of handling large

Figure 1.2 Inside of an Intel semiconductor fabrication facility (Courtesy: Intel Corporation).

wafers and because it takes more processing chemicals to complete the wafer fabrication process. Tool vendors have found a new niche in supplying equipment that can handle these larger wafers. Nevertheless, large wafers have a favourable impact on industry revenues because they yield significantly more devices than smaller wafers. On balance, the shift to larger wafer sizes has had a positive impact and this is a trend that we are likely to see continue for the foreseeable future.

For a successful transition from one so-called technology node to the next a number of process technologies need tweaks and adjustments. Pattern imaging technologies have remained at the forefront of miniaturisation efforts. The application of photolithography to microelectronics made economical production of monolithic chips possible in the nineteen fifties. Since then, advances in photolithography have spearheaded advances in device size down-scaling. Many aspects of the lithography tool have been improved over time. Near field exposure gave way to contact and proximity printing which was later replaced by limited area scanning and the modern step-and-scan approach. This, combined with enhanced optics and illumination systems has made large area patterning with acceptable throughput possible. The continual downscaling of device geometries, however, could not have been achieved without the use of shorter wavelength radiation for pattern exposure. Near-UV radiation at 436 nm (g-line) was replaced by mercury i-line radiation at 365 nm during the early years of commercial chip production. As geometries shrank below a micron, the i-line wavelength too gradually became inadequate as diffraction began to limit pattern resolution. As it gets increasingly difficult to generate large amounts of short wavelength radiation in simple lamp-based atomic discharges so attention was turned towards those lasers that can generate UV radiation efficiently. While a number of laser systems such as the nitrogen laser, argon-ion laser and helium-cadmium laser, among others, can generate ultraviolet radiation, noble gas-based excimer lasers were found to be the most suitable illumination sources for optical lithography. Fluorine chemistry-based excimer lasers generate extremely intense and stable short wavelength UV lines which are very suitable for patterning deep sub-micron features. Krypton fluoride (KrF) lasers that emit at 248 nm were developed first for this purpose. Until recently, KrF lasers were the workhorse of deep sub-micron lithography systems. However, with Critical Dimensions (CDs) getting below 100 nm argon fluoride (ArF) lasers are now the illumination sources in most state-of-the-art lithography tools. The 193 nm radiation from ArF lasers is used to image the mask pattern through a complex system of lenses. The individual elements in these multi-element lens systems are made from materials such as fused quartz, calcium fluoride and lithium fluoride – these materials are some of the only ones available that are transparent to wavelengths below 200 nm. Calcium fluoride probably holds the most promise for future lithography systems because of its high laser damage threshold – an essential attribute for a lens material that has to pass large amounts of energy to expose relatively insensitive short wavelength resists. Other materials tend to develop material density and refractive index fluctuations at high radiation fluences. The complex nature of modern lithography lenses means that these highly engineered units are very bulky, expensive and require periodic maintenance. Figure 1.3 here shows a commercial lithography lens system for use with excimer laser systems.

Figure 1.3 A photolithography projection lens assembly (Courtesy: ASML Corporation).

With shrinking feature sizes, the 193 nm ArF radiation will also prove inadequate and tools will transition to 157 nm wavelength radiation available from Fluorine excimer laser sources. Beyond that, it would no longer be possible to build an entirely refractive i.e. lens-based imaging system. Lithography tools in the future will most probably use catadioptric i.e. lens-mirror combination systems or even completely reflective systems. These systems have the additional advantage of low chromatic aberration that can accommodate the natural line width of excimer lasers. Unfortunately, to image with shorter wavelengths one also needs to use resists that are sensitive at these wavelengths. Special deep-UV photoresists have been developed for photolithography at shorter wavelengths. There are challenges here as well. The novolac and poly(hydroxystyrene)-based 'conventional' resists that are used at 365 nm and also at 248 nm wavelengths become increasingly opaque at shorter wavelengths. With absorption lengths decreasing to just around 40 nm at 193 nm these resists are not of much use for exposures with ArF radiation. Methacrylate-based resists are fairly transmissive at such short wavelengths and several methacrylate-based photoresists have been developed that conform to the thermal and transparency requirements of 193 nm exposure processes. In fact, being also sensitive to 248 nm radiation, the methacrylate-based resists have been used as dual wavelength resists in mix-and-match 248 nm/193 nm lithography systems. Their only significant disadvantage seems to be their relative lack of etch resistance in dry etch processes. This issue becomes quite severe for chlorine-containing chemistries, for instance those based on BCl_3, that are used for metal etching. Their etch resistance can be increased by incorporating high molecular weight carbon-rich polymers with them. These can be alicyclic polymers but not aromatic polymers as the latter have very low transparency below 200 nm. Multi-layer and top imaging resists are other technologies that are under investigation for lithography at 193 nm and shorter wavelengths. Several experimental top surface imaging resists have shown excellent performance in the laboratory. Their main advantages seem to be larger depth of focus and the elimination of antireflective coating. The chemistry of short wavelength 193 nm and post-193 nm resists remains an active topic of research in several industrial, academic and consortium laboratories around the world.

Leading technology it may be but reduction in exposure wavelength is not the only refinement that has been paving the way for smaller feature sizes in chip manufacturing. Photo-mask manufacturing itself has undergone significant changes over the years. Commercial reticles are manufactured on quartz or fused silica blanks that are transparent to the exposure wavelength. Both laser raster writing and electron beam writing are in use for mask manufacture. In recent years, increasing use has been made of various mask-level optical techniques for accurate image reproduction. These resolution enhancement techniques include phase shift masks and optical proximity correction. The former uses special phase shifting materials to create an interference pattern at the image plane. The two most developed variants of this technique are attenuated phase-shifting and alternating phase-shifting masks. Attenuated phase-shifting makes use of optical phase-shifting materials such as molybdenum silicide (MoSi) in selected areas of the reticle. This material allows around 10% of the incident UV radiation to pass through it but, with appropriate choice of MoSi thickness, the light that does get through is shifted in phase by 180° as compared to light that simply passes through clear quartz. As shown in Figure 1.4 here for the attenuated phase-shifting approach, the resulting slight exposure of the resist is insufficient to form a permanent image after development because it is considerably less than its exposure threshold. The 180° phase contrast between clear and MoSi-shifted areas produces darker areas where needed than would be the case with a simple binary mask where diffraction of light between closely located features causes a build-up of considerable light intensity underneath chrome areas. Thus, as far as the resist is concerned, it experiences larger contrasts between light and dark areas with phase-shifted aerial illumination than with binary light/dark patterns, allowing much finer pattern details to be faithfully reproduced in the resist image.

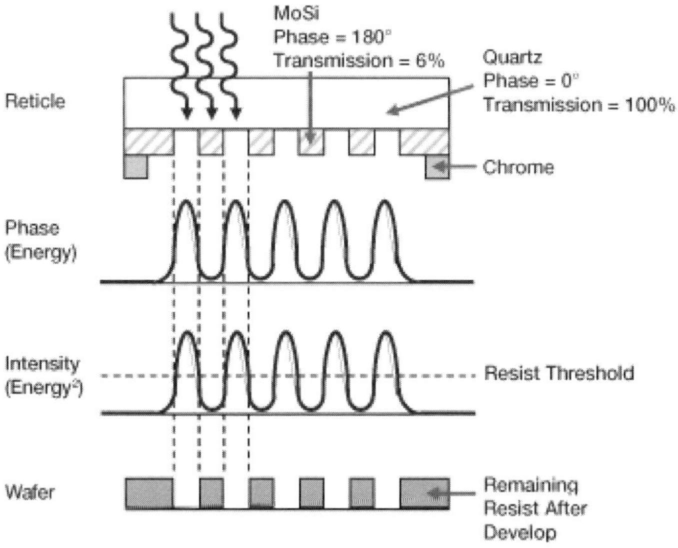

Figure 1.4 Operation of an attenuated phase-shift mask (Courtesy: ASML Corporation).

Alternating phase-shifting mask technology is somewhat more sophisticated as well as expensive to implement. Here, reticle quartz is etched in selected locations to produce 180° shift in phase of the transmitted UV light as compared to adjacent areas of un-etched quartz. The resulting interference pattern creates the desired aerial image at the resist plane. As an example, as seen in Figure 1.5, chrome lines on the reticle are bordered on one side by quartz of phase 0°, and on the other side by quartz of phase 180°. As the phase goes from positive to negative, it passes through 0°. The intensity (proportional to the square of the phase) also goes through 0, making a very dark and sharp line on the wafer.

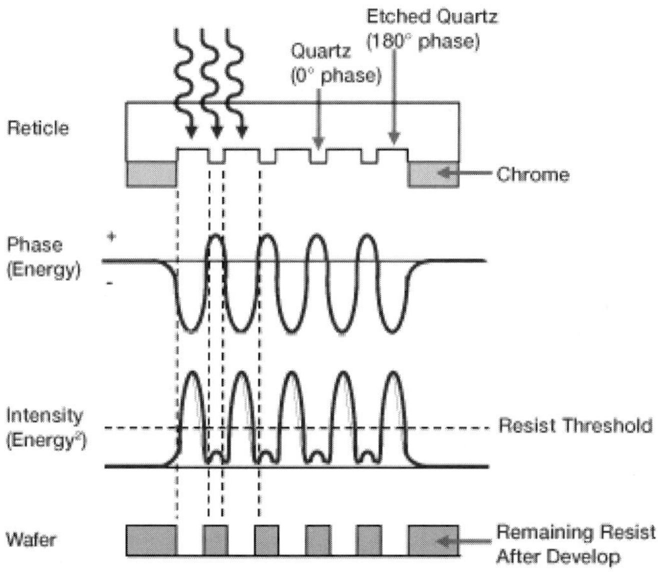

Figure 1.5 Operation of an alternating phase-shift mask (Courtesy: ASML Corporation).

While phase-shifting requires special materials and manufacturing techniques, Optical Proximity Correction (OPC) relies on simply modifying the pattern on the reticle in order to produce finer circuit features. The main idea is that unwanted diffraction effects could be compensated to a large extent by introducing extra features on the mask plate such that the combination produces the desired aerial image at the resist plane. OPC is most effective in accounting for photolithography process variations due to differing feature densities within a die image. This is common in today's mixed-signal and memory-logic combination chips. Proximity correction features such as scatter bars and 'serifs' added to line ends enable increase in lithography process latitude without resorting to more complex techniques. A similar technique is the selective addition of light absorbing pixels to compensate for CD variations within an exposure field. This relies on laser written pixels placed on the reticle at clear apertures so that some light is absorbed while passing through them. The exposure underneath these locations is thus reduced which, in turn, shortens CDs in such a manner that after exposure all similar features begin to come out with the same size. This technique has been found to be very useful for tightening specifications on intrafield CD variations.

Today's commercial photolithography systems generally utilize all resolution enhancement techniques described above. However, with continually shrinking device geometries new, and sometimes radical, breakthroughs in technology are needed in order to continue moving on the semiconductor industry roadmap. One such technology is the recent development of liquid immersion lithography as a mainstream deep UV optical lithography technique. Here we need to take a look at how imaging resolution is related to other optical parameters. The Rayleigh equation determines the resolution or the minimum CD that can be printed with a given imaging system. It relates the minimum printable feature size to the parameters of an optical lithography system, as:

$$W = k_1 \lambda / NA$$

where, k_1 is the resolution factor, λ is the wavelength of the exposing radiation and NA is the numerical aperture of the imaging system. Thus, printable feature size W could be reduced not only by reducing the exposure wavelength λ but also by reducing the k_1 factor and by increasing the NA of the lithography system. k_1 is dependent on several factors such as wafer planarity, resist characteristics, use of OPC techniques etc. Improvements in all these fields have pushed k_1 down to values of around 0.3 and 0.25 may be achievable in the future. This leaves us with the numerical aperture of the imaging system. This last parameter is the product of the refractive index of the surrounding medium with the sine of the light acceptance angle of the lens, as:

$$NA = n \sin\theta$$

Obviously with $\sin\theta \leq 1$ and $n_{air} = 1$, the NA of any lens system will remain less than unity. With design advances in optical lithography systems the numerical aperture of the lens train has gradually increased from about 0.4 to 0.85 that is typical of today's systems. With practical lens systems, numerical apertures of up to 0.93 may be achievable but that is about as much as could be realised with conventional systems. If the NA could be pushed beyond unity then dramatic improvements in lithographic resolution become possible and that is exactly what immersion lithography can do. Increase in the depth of focus is an additional advantage of immersion systems. This new technology is so promising that many academic and industrial labs have been busy developing it for the past several years. Immersion lithography is based on increasing the numerical aperture of the imaging system by immersing the projection lens in a refractive liquid. In practice, a liquid, transparent to the exposure radiation, is interposed between the final projection lens and the resist covered wafer. Obviously, the immersion fluid must be non-invasive to the resist material and the lens when in contact with them for short periods of time. Ultra-pure semiconductor grade water is currently the favoured immersion fluid. Filling the gap between the lens and the wafer with water increases the resolution by approximately a factor equal to the refractive index of water. Numerical apertures of up to 1.4 can be attained with this approach and the use of special high refractive index immersion fluids might permit NA extension to beyond 1.5. Immersion technology will certainly enable production down to the 32 nm node and possibly beyond. However, implementing immersion technology was no easy task as there were a number of technical challenges that had to be met before it became a commercially viable technology. These included the detailed mechanics of dispensing and removing the immersion fluid and keeping it in place as the wafer stage undergoes large accelerations and decelerations during the operation of the scanner. Most of these issues have now been satisfactorily addressed so that 193 nm immersion lithography systems are now commercially available from a number of manufacturers.

A well-designed lithographic process, utilizing an appropriate combination of the technologies described above, should be capable of sustaining wafer patterning down to the 32 nm node. We are further assured of this because beginning from the 65 nm node a double patterning approach has been of help in successfully meeting photolithographic targets for patterning both advanced FETs and memory cells. Double patterning, which can be used in a number of different ways, is essentially performed by exposing two successive photoresist layers using two different photomasks so as to achieve an increase in feature density by up to a factor of two. Depending on the implementation, there may be one or more material deposition and etching steps in between the two photolithographic exposures. Double patterning, and other similar multiple patterning techniques, cannot increase the optical resolution of the underlying lithography process but it can still achieve higher pattern density through its geometric scaling approach.

Beyond the 32 nm technology node photolithography may have to face very tough challenges that may not be answered by using extensions of resolution enhancement technologies known today. Exposure wavelengths may have to be further reduced to allow optical patterning of circuit features. Various academic and industrial groups are looking at Extreme Ultra Violet (EUV) lithography technologies as a way out of potential problems that the industry is likely to face in the second decade of this century. Both very deep UV radiation sources and associated optics are under development. At first, there was much interest in synchrotron X-ray sources for generating extremely short wavelength exposure radiation but in recent years smaller and less complicated sources have found more favour and several research groups are looking at ways of obtaining useful amounts of radiation in the tens of nanometer region from tabletop-size systems. An experimental 13.5 nm system under consideration by the Extreme Ultraviolet Lithography System Development Association is shown in Figure 1.6. A Nd:YAG laser illuminates a cold xenon jet with its 1064 nm infrared emission. This creates a Xe plasma containing electrons and variously charged positive ions of xenon. Electronic transitions in heavy singly and multiply charged xenon ions produce strong emissions in the soft X-ray region around 13.5 nm. Using a reflective system this radiation is focused on the wafer through a special reticle. Of course, very precise stage positioning systems are also required to benefit from the fine feature capability. Field-to-field stitching tolerances in the range of 5 to 10 nm are required. This level of control is only possible through interferometric systems. This is one aspect of the EUV lithography package which is already fully developed and will require only minor adjustments. Over the years, masks and reticles that can work effectively at such short wavelengths have posed a very considerable challenge. Ultra-thin silicon nitride masks with molybdenum/silicon multilayer radiation blocking coatings have shown much promise but these remain plagued by a number of problems including fragility, manufacturing expense and difficulty in defect repair. Luckily, this area has been recognized for quite some time as one that will require considerable development and has seen substantial advances. Very specialized resists will also need to be developed to work with far UV systems; some of which may work through entirely novel routes. Hard X-Ray and ion-projection lithography systems were also considered as potential imaging tools but at this time EUV projection lithography remains the favourite next generation lithography technology.

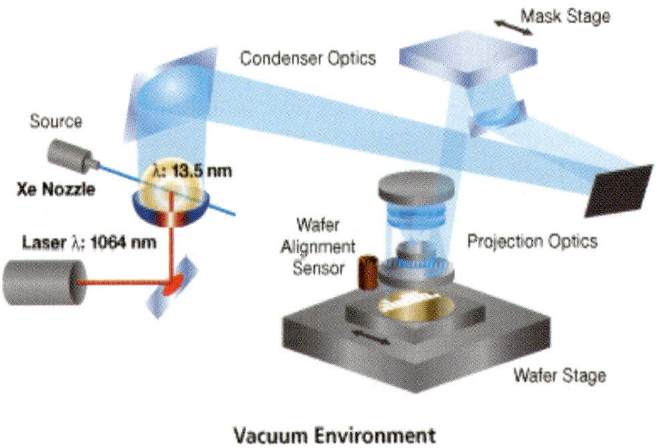

Figure 1.6 An experimental EUV lithography system.

Circuit patterns carved in resist on semiconductor wafers need to be transferred onto the wafer material as permanent circuit features. In additive pattern transfer, a material is usually deposited as a thin film on the patterned resist through a physical deposition process such as metal evaporation or sputtering. It is essential that this process doesn't heat up the substrate material because at this stage it is covered with a polymeric resist layer. After thin film material deposition the entire wafer is submerged in a resist solvent that dissolves away the resist and any material lying on top of it. Deposited material is then left on the substrate where there were openings in the patterned resist. Though straightforward in concept, this form of additive or lift-off patterning is not used in the silicon chip industry to any reasonable extent because of process reliability concerns. The other pattern transfer process – subtractive patterning is the prevalent process in contemporary IC fabrication technologies. For subtractive pattern transfer the material to be patterned is first blanket deposited on the substrate and then the required pattern is defined in resist coated on top of the material. Next, etching is performed to selectively remove material where it is exposed through openings in the resist. Afterwards, the resist is removed and the patterned material is left behind. Traditionally, etching was carried out with various chemical solutions that could selectively etch through the material being patterned but not affect the underlying substrate. The major problem with wet etching, its isotropic etching characteristic, becomes increasingly evident as circuit features shrink below ten microns. Etching proceeds at the same rate both vertically downwards and laterally side-ways. In amorphous materials the rate of vertical and lateral etchings are the same but there are directional dependencies in etching crystalline materials. Lateral etching is generally undesirable as it gradually erodes lateral pattern features and degrades spatial definition of circuit patterns. For IC critical dimensions less than about ten microns wet etching is of little or no use as a patterning process. Fortunately, at the time that this could have become a serious issue impeding further development of integrated circuit technology, the now familiar dry or plasma etching technology was developed to solve this problem. While there are many variants of dry etching technology all of them use reactive plasmas to slowly erode away material. Plasma etching is without doubt the most important enabling technology after advanced photolithography that has made possible the present generation of deep submicron ICs. Dry etching is a so-called synergistic process; the physical action of ion bombardment on the exposed surface is supplemented by chemical reactions between the reactive species found in the plasma and the surface material. Both of these mechanisms operate together to remove material in the dry etching process. The synergy between physics and chemistry provides both efficient etching and allows great control over etching characteristics as either the physical or the chemical process could be enhanced in order to confer desired properties on the etching process. Thus, for example, increasing the ion bombardment by increasing the bias voltage in the plasma chamber causes the etching to produce a more vertical profile with little lateral undercutting. This flexibility in controlling etch side-wall profile has also been made use of in creating a variety of structures for micro- and nano-electromechanical systems. The chemical reactions inherent in plasma processing enable even extremely hard materials such as diamond, tungsten carbide and palladium to be etched with ease. Without the ability to etch materials like these modern semiconductor industry would not have been possible. Dry etching techniques continue to be developed further and new variants are frequently introduced by researchers and tool makers to tackle increasingly complex etching tasks. Figure 1.7 shows a row of dry etching tools in a fab setting with operators loading 25-wafer pods.

Figure 1.7 Dry etch tools in a fab environment (Courtesy: Intel Corporation).

Some techniques, primarily developed for dry etching, have also migrated to material deposition. This is possible because a low temperature material deposition technology relies on gaseous discharges instead of heat to provide the energy for reactant molecules to dissociate. Plasma Enhanced Chemical Vapour Deposition (PECVD) deposits dielectric materials such as silicon dioxide and silicon nitride through a low temperature plasma-driven process. Techniques such as Remote Plasmas (RP) and Inductively Coupled Plasmas (ICP) originally developed for reactive ion etching processes have migrated to PECVD tools with beneficial effects.

With increasing sophistication of on-chip circuitry, comprised of millions of active switching devices, the interconnection patterns required to connect them have also become incredibly complex. In order to accommodate the wiring complexity more than one level of wiring plane is required. Two to four levels of interconnections are quite common for analogue and medium complexity digital ICs. State-of-the-art microcontroller chips, however, require no less than six levels of wiring. As the number of wiring layers increases, the planarity of these layers decreases. The undulating topography begins to affect photolithography process latitudes. This is because high resolution lithography, especially when printing over large areas, has a very small depth of focus. Thus when locations on a wafer protrude above the optimum imaging plane or go below it by more than a few nanometres then resist in those areas doesn't get properly exposed. Before the era of sub-micron feature integrated circuits the way around this difficulty was to planarise inter-metal dielectric layers through a process that consisted of the deposition and subsequent melting of low melting point glass over-layers. The re-melted glass provided a flat layer on which lithographic exposure could be performed satisfactorily. These reflow processes worked adequately well until the temperatures required to cause glass overflow became incompatible with the decreasing thermal budgets of increasingly complex chips. For some time, this was accommodated by the use of boron and phosphorus-containing glasses that have a lower glass transition temperature than pure silica glass. A point was reached when even these boro-phosphate glasses couldn't provide the low temperatures reflow capabilities needed to remain compatible with the rest of the fabrication process flow. A radically new planarisation technology was needed and this appeared in the form of Chemical Mechanical Planarisation (CMP). This process operates by polishing the surface of a rotating wafer with counter-rotating felt pads soaked with especially loaded slurries. Like dry etching, CMP too is a synergistic process where both chemical and physical abrasion effects are involved in removing material from the surface of a wafer. Unlike dry etching, however, CMP operates to remove material from all over the wafer with protruding points experiencing a faster removal rate than other locations. This way, the wafer gets increasingly planarised with the passage of time. Different slurry mixes and different slurry concentrations could be used to customise the process to satisfy different planarisation requirements. Slurry mixes often contain an oxidising agent to oxidise the material that needs to be removed and an abrasive agent, such as silica particles, to remove the oxidised material through a mechanical grinding action. With proper wafer balancing on the CMP platen, even dies close to the wafer edge can be planarised thus lifting the overall yield. Being a 'dirty' process, CMP is usually carried out in a different part of the fab than the rest of the wafer foundry. A thorough cleaning is needed after the process in order to remove all traces of slurry particles and abraded wafer material. This process is capable of achieving astonishing across-the-wafer planarisation – values of 10 nanometre RMS flatness being quite common. The following process could then start from a flat wafer – with consequent gains in photolithography latitude. CMP is one example of a radical process that has made complex IC fabrication possible. Figure 1.8 here shows a planarised wafer ready for subsequent processing.

Figure 1.8 Mirror-like surface of a post-CMP wafer.

Advances in virtually all aspects of chip manufacturing has helped keep the semiconductor industry remain profitable over the years. While lithography, dry etching and CMP have received the most coverage, other processes – both fabrication and metrological had to be tweaked to keep to the semiconductor industry roadmap. There have been advances in ion implantation, rapid thermal processing, deposition of dielectric layers, monitoring of on-wafer particulates, wet cleaning of sub-micron features and measurement of ultra-thin film thicknesses, to name a few. Most of these enhanced processes are developed by tool manufacturers and supplied to chip manufacturers as comprehensive tool/process packages. Major foundry-owning chip manufacturers do perform their own technology development research, however, much of that is focused at process integration rather than the development of individual process modules.

While device fabrication tools and techniques play central roles in the production of commercial semiconductor chips, the industry will not be able to function without the existence of equally complex process monitoring and diagnostic tools. In order to maintain acceptable die yields, the many dozens of individual processes that constitute an IC process flow have to be diligently monitored. Typically, every major process is paired with a suitable process characterization step. Thus, for example, etches may be monitored by Atomic Force Microscopy (AFM) and thermal silicon oxide growth may be monitored by ellipsometry. There are also machine vision-based optical inspection and scanning electron microscopy-based inspection tools that verify pattern fidelity after each lithography step to make sure that all expected features are in their right places. Typically, such tools are loaded with a master pattern file and acceptable tolerance information and then they scan either all wafers passing through certain processing steps or a selection of random wafers. Departure from statistical tolerance levels is then flagged by the equipment and is transmitted to production control systems. Similar tools are also employed to monitor the distribution of particulate debris emanating from etching and deposition processes. The resolution of these systems has increased dramatically in recent years to keep pace with miniaturization trends. Results from such inspections come in useful when chip yields are being improved or process failures are being addressed. In many cases, timely intervention can avert a potentially catastrophic manufacturing problem. Figure 1.9 shows a multifunction optical metrology tool from Zygo Corporation that can measure critical

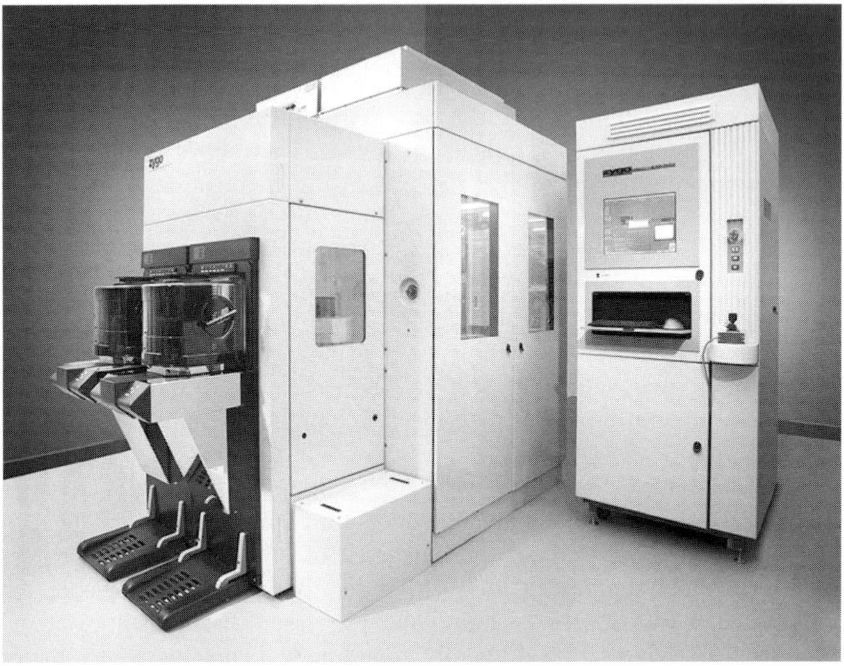

Figure 1.9 A multifunction optical metrology tool (Courtesy: Zygo Corporation).

3-dimensional features on production wafers for highly reliable process control.

In the next few years we are going to see further advances in miniaturization. These will be driven both by improvements in existing processes and by the development of radical new technologies. When combined with better circuit designs and the use of such new technologies as chip stacking, we shall witness new generation of increasingly sophisticated ICs whose power could not even be imagined a few years ago.

By the third decade of this century semiconductor industry as we know it today will no longer exist. Its demise will be brought about by fundamental limitations of the same physics and materials science that have been so successfully exploited throughout its evolution. Once device dimensions get around 10 nm the physics of their operation changes drastically. Ballistic carrier transport and quantum mechanical effects such as carrier tunneling become increasingly important.

Statistical fluctuations in parameters such as channel doping – unimportant for larger devices – assume increased significance. Current leakage through extremely thin gate dielectrics begin to seriously affect device performance. These and other effects will combine to herald the end of the Complementary Metal Oxide Semiconductor (CMOS) era in its present form. Novel device architectures and fundamentally new device operating principles will have to be explored for the post-CMOS era. While we don't know what form these new technologies will take, we do know that at least some of them will be the derivatives of present day research into nanostructures – both electronic and photonic. Contemporary nanotechnology research is throwing a lot of light on the behaviour of both matter and structures at the smallest scales possible. Science carried out at the nanometer scale is fundamental because here we come face to face with the graininess of matter itself. Structures smaller than 100 nm in any dimension possess only a few dozen atoms in that dimension and can display behaviour that is far removed from what is seen in larger structures. A lot of what is going on in current nanotechnology research is the investigation of this novel behaviour. On another front, new nanometer scale fabrication technologies are being developed. Some like advanced photolithography are extensions of well established techniques whereas others such as nanoimprint lithography are relatively recent developments. Both fundamental research at the nano scale and the development of nanoscale fabrication techniques are essential for exploring possibilities for future technology choices.

Matter can be structured on the very small scale by a number of techniques that depend on the particular direction in which the structuring takes place. Vertical layers can be very precisely placed by a variety of epitaxial and non-epitaxial growth techniques. Both Chemical Vapour Deposition (CVD) and Physical Vapour Deposition (PVD) processes are employed for this purpose. Molecular beam techniques such as Molecular Beam Epitaxy (MBE) are especially powerful non-equilibrium growth techniques that are capable of producing epitaxial growth on appropriate substrates to the accuracy of just a single monolayer of atoms. Its CVD analogue – Metal Organic Chemical Vapour Deposition (MOCVD) which also goes by the name of Metal Organic Molecular Beam Epitaxy (MOMBE) is now the standard industrial process for growing layered compound semiconductor structures for devices such as semiconductor lasers, LEDs and transistors. Computer controlled growth sequencing in MBE and

MOCVD allows very complex heterostructures to be grown with very high repeatability. Lateral structures are not as easy to form and require elaborate lithographic processes. While high resolution photolithography, as described earlier, is capable of routinely patterning sub-100 nm features the most widely used technique for research purposes is electron beam lithography, also referred to as e-beam lithography.

Electron beam lithography that was developed in the nineteen fifties and sixties into an extremely high resolution patterning technique has remained the ultimate nano-patterning technology. Its conceptual simplicity combined with impressive capability has made it the favourite of researchers all over the world. The technique itself developed directly from Scanning Electron Microscope (SEM) technology. Thus it uses much of the same hardware and indeed SEMs could be modified fairly easily to convert them into e-beam lithography tools. The main modification that is needed is the addition of a beam blanker to the electron beam column that can switch the beam on or off as needed. Figure 1.10 shows an electron beam lithography system within a clean room environment.

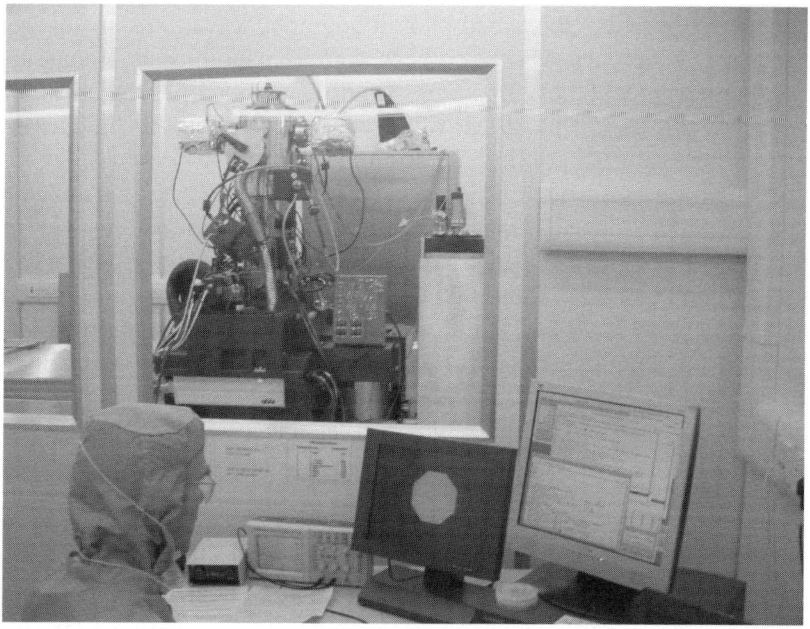

Figure 1.10 An electron beam lithography system inside a clean room environment.

The idea behind electron beam lithography is quite simple: a beam of electrons accelerated to around 50 keV is made to impinge on a resist coated sample held on a stage in vacuum. The special resist responds to the electron bombardment in one of two ways. Either the resist gets cleared from the exposed locations after a subsequent development step or it is cleared from areas that were not exposed to electrons. The former are positive tone resists whereas the latter are negative tone resists. Polymethyl Methacrylate (PMMA) is the most widely used positive tone e-beam resist. Electron bombardment causes PMMA molecular chains to fragment into shorter chains in a process called 'chain scission'. The shorter chains are very much more soluble in organic solvents used as developers. Methyl isobutyl ketone (MIBK) is a widely used developer (in practice it is usually diluted with isopropyl alcohol). It should be mentioned here that at very high dosage of electron beam exposure PMMA cross-links to form an extremely high molecular weight resin that is almost completely insoluble in organic solvents. Thus at high exposure doses PMMA acts as a negative tone resist. However, as the cross-linked resist gels when soaked in solvents so used in this fashion PMMA provides much inferior resolution and is, therefore, seldom used as a negative tone resist. Until recently PMMA was the highest resolution positive tone resist known. Its sensitivity depends on its molecular weight distribution with high molecular weight dominant samples being less sensitive than low molecular weight samples. This and other e-beam resists are coated onto substrates using a spin coater and then baked at around 90°C to remove any solvent trapped in the resist film. During the lithography process PMMA is exposed not only by the incident electron beam but also by electrons that are backscattered from the substrate. The higher the mean atomic weight of the substrate material the higher is this secondary exposure which serves to spoil the achievable resolution somewhat. Very closely spaced features are the ones that are most affected by this 'proximity effect'. Both experimental techniques and electron beam dose correction algorithms have been developed to reduce the impact of proximity effect on lithographed patterns. Higher electron beam accelerating voltages move the maximum scattering zone to deeper inside the substrate and thus reduce proximity effect. Thin resist and substrate layers, where practicable, are also effective in reducing proximity exposure. Modern state-of-the-art e-beam lithography tools can write structures less than 10 nm wide and several microns long over a one square millimeter field. A commercial electron beam lithography system capable of directly writing features narrower than 10 nm is shown in Figure 1.11 here.

Figure 1.11 A high resolution electron beam lithography system (Courtesy: Raith GmbH).

The dynamics of electron beam scattering leads to a pattern of resist exposure that results in an undercut profile. This is opposite to what happens in photolithography. An undercut resist profile makes it easier to pattern material through additive i.e. post lithography blanket deposition processes. This is a very significant advantage of e-beam lithography that enables the fabrication of ultra-small metal structures. On the other hand, PMMA does not perform so well when it comes to subtractive patterning processes because of its rather limited resistance to dry etch ambients. Acrylic polymers are easily attacked by reactive species in dry etching plasmas. As stated earlier, photoresists based on novolac resins that are based on aromatic hydrocarbons are comparatively much more resistant to dry etching plasmas. Presence of aromatic ring groups confers dry etch resistance to resists, whether optical or electron beam. Another class of resists that is now widely used for e-beam lithography is based on poly(silsesquioxane) organo-silicon materials such as Hydrogen SilsesQuioxane (HSQ). In contrast to acrylic-based resists such as PMMA, this is a negative tone resist. On electron beam exposure and subsequent thermal treatment HSQ changes from a cage-network combination structure to a completely network permeated structure of Si-O bonds. This silicate material is hard and resistant to many chemicals and consequently can be used as a good hard mask. Modern HSQ resists have resolutions that are comparable to that of PMMA. Because these resists also have good dry etch resistance so their use is growing rapidly and within a few years they have replaced conventional acrylic resists for many patterning processes.

While electron beam lithography is excellent for making one or a few prototype devices it is not a suitable production tool for volume manufacture. The benefits of e-beam lithography could be retained in a commercial production environment through the use of contact pressure stamping tools in a technique called Nano Imprint Lithography (NIL). At first, the idea is similar to that of photolithography in that a volume manufacturing patterning tool is made using e-beam lithography. However, instead of an image in chrome, NIL makes use of a special stamp that carries structures to be patterned as an image in relief. Once such a stamp has been satisfactorily fabricated it could be used to stamp patterns on special thermoplastic or thermosetting resins. The topographical image in resist could subsequently be transferred to the substrate using dry etch processes. Not being an optical technique, NIL avoids the use of any radiation for imaging, although some variants do use UV radiation for resist curing. Nano imprint lithography has been shown to be capable of deep sub-100 nm resolution and is a serious contender for the commercial production of a variety of nanostructures in the near future. Nano imprinting is performed in several ways. An NIL stamp in a hard material such as diamond-like carbon, quartz or silicon carbide could be pressed on to a resist-covered substrate. Under the influence of heat and pressure the resist cures and takes the inverse shape of the NIL stamp. In another technique, a stamp is made on a UV transparent material such as quartz and impressed on to a UV-curable resist coated on top of a substrate. The arrangement is then illuminated with UV radiation that passes through the stamp and cures the polymer into a hardened, cross-linked state. All NIL processes create a relief image in speciality NIL resists. This image is subsequently transferred to the substrate or a hard mask through a dry etch process.

Another optical lithographic technique that has found use in the replication of strictly periodic patterns is interferometric lithography. The idea here is to create a periodic pattern of electromagnetic standing waves by making two coherent light beams, usually from a laser source, interfere with each other. Coherency ensures that a stable fringe pattern is formed. If the process is carried out on a photosensitive resist then the pattern of interference fringes could be recorded in the resist material as a real image. After development, this image could be transferred to the substrate using usual dry etch processes.

While lithographic methods can be used to sculpt matter precisely these begin to lose steam as patterned structures get smaller and smaller. Lithography also gets progressively less useful as the number of structures to be produced increases to billions or more at a time. When occasion demands the replication of exceedingly large numbers of very small structures then use is made of the so-called self-organization or self-assembly techniques. These get their remarkable power from stereochemical effects that are evident on the surfaces of textured substrates and specifically-shaped molecules. DNA molecules have been frequently used to compose intricate structures in the recent past and other similar giant molecules have also been investigated in this role. Self-organized structures are naturally less perfect than lithographed structures but in situations where large numbers of identical structures is the main requirement self-organization proves to be a very viable and powerful technique. This is mainly because in such cases the imperfections of individual structural elements get statistically evened out. Often a combination of lithography and self-assembly can produce desired structures that would be impossible to fabricate using just lithography or self-assembly alone.

A combination of the nanostructuring processes and characterization described above together with other similar fabrication technologies will lead to advances in nanotechnology that will form the basis of tomorrow's commercial technologies. The rest of this book is comprised of a selection of chapters adapted from published papers on nanotechnology by a varied group of scientists and engineers in both industry and the academia. These chapters are divided into two main groups. The first set of nine chapters is mainly focused on applications where electrons play a significant role. Whereas the next set of six chapters deals mainly with optical themes. This collection is representative of contemporary research in nanostructure science and technology and illustrates the wide range of materials and processing techniques that are under development for future applications in electronic and photonic technologies.

2

ASSEMBLING FERROMAGNETIC SINGLE-ELECTRON TRANSISTORS WITH ATOMIC FORCE MICROSCOPY

Håkan Pettersson, Ruisheng Liu, Dmitry Suyatin and Lars Samuelson

Ferromagnetic Single Electron Transistors (F-SETs) comprise ferromagnetic electrodes connected to a ferromagnetic- or non-magnetic central island via tunnel barriers. These devices are important for studies of spin-transport physics in confined structures. Here we describe the development of a novel type of AFM-assembled nano-scale F-SETs suitable for spin-transport investigations at temperatures above 4.2 K. The ingenious fabrication technique means that their electrical characteristics can be tuned in real-time during the fabrication sequence by re-positioning the central island with Ångström precision.

2.1 Introduction

Single-electron transistors (SETs) have attracted extensive interest both experimentally and theoretically because of their possible use as very sensitive electrometers, as well as for investigations of fundamental physics phenomena[1-3]. The basic SET design includes drain and source electrodes connected to a central island by tunnel barriers. In addition, a gate electrode is capacitively coupled to the central island. Figure 2.1 shows the essential parts of the SET. The central island is so small that a charging energy given by $E_c = \frac{e^2}{2C}$, where e is the electron charge and C is the capacitance of the island, is required to add one single electron to the island. If the thermal energy $k_B T$ is smaller than this charging energy, the electron transport is blocked below a certain drain-source bias voltage. This phenomenon, known as Coulomb blockade, is a fundamental property of single-electron transistors[4].

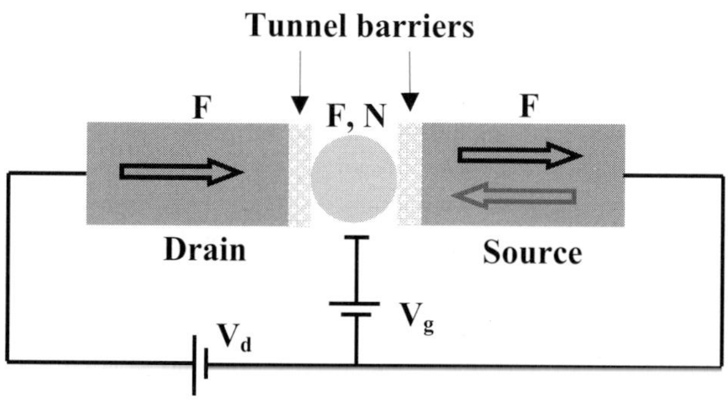

Figure 2.1 Schematic figure of a ferromagnetic single-electron transistor (F-SET). The letters "F" and "N" indicate a ferromagnetic metal and a non-magnetic metal, respectively. The arrows denote relevant magnetization orientations of the electrodes.

A second condition for the observation of charging effects is that quantum fluctuations of the number of electrons on the island must be negligible. To ensure that the electrons are well localized on the island, tunnel barriers with resistances $R_T > \frac{h}{e^2} = 25.8\ k\Omega$ are needed to isolate the island from the electrodes.

More recently, it was found that the interplay between Coulomb blockade in ferromagnetic single-electron transistors (F-SETs) and spin-dependent tunneling processes can give rise to novel phenomena, such as enhanced tunneling magnetoresistance (TMR) and TMR oscillations with applied bias[5,6]. The tunneling magnetoresistance is defined as

$TMR = \frac{R_{AP} - R_P}{R_P}$, where R_{AP} and R_P are the device resistances in the antiparallel and parallel magnetization configuration, respectively. Enhanced TMR is very promising for applications e.g. magnetic random access memories (MRAM), for read/write heads in hard discs as well as in various other spintronic devices[7]. An important prerequisite for building spintronic devices is that the orientation of the spin can be controlled over a length scale that is typically larger than the size of the device. There is thus a great interest in understanding of the different spin-flip mechanisms active in nano-scaled magnetic devices.

The most important mechanism behind spin relaxation in metals is believed to be the spin-flip scattering of electrons due to interactions with phonons and impurities, as suggested by Elliott[8] and Yafet[9]. Because of the spin-orbit interactions produced by lattice ions, electronic Bloch states contain both spin up and spin down components. The states can still be polarized by a magnetic field (so we can label them spin up and spin down) but because of the spin mixing, even a spin-independent perturbation induced by phonons or impurities leads to a degrading of a spin-polarized electron population. It is important to note that it is the spin-orbit interaction in conjunction with phonons and impurities that induces spin relaxation, not the spin-orbit interaction alone.

Interestingly, it has been observed that the spin relaxation time can be enhanced by orders of magnitude as the size of the device is drastically reduced to merely a few nanometers[10]. A very promising type of device for studying spin-transport physics in confined structures is ferromagnetic single-electron transistors, F-SETs (shown in Fig. 2.1). In F-SETs, ferromagnetic electrodes are connected to a ferromagnetic- or non-magnetic central island via tunnel barriers. To efficiently inject spin-polarized electrons into the central island, the two electrodes are normally fabricated with different shapes or different ferromagnetic materials (e.g. Ni, Co, Fe), so that their magnetic moments can be changed from a parallel to antiparallel configuration by means of an external magnetic field (indicated by red and black arrows in Fig. 2.1).

So far, most experimental reports on TMR in F-SETs deal with relatively large devices where a sample temperature in the range of a few tens ~ hundreds of mK is required in order to observe any significant TMR[5-6, 11-12]. The existing reports dealing with investigations of TMR and spin-accumulation in ultra-small SETs, where the central island presumably is a grain with a size of merely a few nanometers, have the limitation of a poor control of the size, shape and even composition of the central island[10].

In this chapter, we describe the development of a novel type of AFM-assembled nano-scale F-SETs suitable for spin-transport investigations at temperatures above 4.2 K[13]. With this technique, SETs with island sizes down to 10 nm can be realized. Further advantages include well-controlled islands in terms of size, shape and composition as well as electrical characteristics that can be tuned in real-time during the fabrication by re-positioning the central island with Ångström precision. Compared to previous AFM-manipulated SETs[14], our design differs significantly in that the electrodes are ferromagnetic, that stable insulating NiO tunnel barriers are fabricated in contrast to air gaps, and that a single 30 nm island is used as a central island rather than several bridging 50 nm discs. The fact that a single island is used is crucial for spin-injection experiments. In addition, the device fabrication complexity is reduced and the design is more robust with better long-term stability.

2.2 Experimental Details

The devices were fabricated on top of a 100 nm thick SiO_2 layer grown on a Si substrate. A layer of 950 PMMA was spun on the substrate at 6000 rpm for 30 seconds, after which the sample was baked in an oven at 180°C for 1 hour. A grid of Au discs, 30 nm in diameter and 30 nm thick, was prepared by conventional electron-beam lithography followed by thermal evaporation and subsequent lift-off (shown in Figure 2.2a). The metals were normally evaporated at a pressure of $\sim 5\times10^{-7}$ mbar. Ferromagnetic Ni source and drain electrodes, 30 nm in thickness, were formed in subsequent similar process steps together with a 30 nm thick Ni side-gate. The width (80 nm) of the drain electrode was comparable to the size of single Ni domains[15, 16], while the source electrode was considerably wider (220 nm).

Because of this shape anisotropy, the two electrodes are expected to undergo magnetic reversal at different magnetic field, confirmed by performing micromagnetic simulations using the OOMMF code[17] which gave the coercivity field of 40 mT and 90 mT for the source and drain electrodes, respectively. By sweeping the magnetic field it is thus possible to switch from parallel to antiparallel alignment of the magnetic moments of the two electrodes. Tunnel barriers of NiO were fabricated by sequential Ar plasma etching at a pressure of about 1.0×10^{-4} mbar for 5 minutes and O_2 plasma etching at a pressure of about 1.0×10^{-1} mbar for 3 minutes. Following these process steps, the sample was mounted on a standard chip carrier, bonded and placed in the AFM which was operated in air at room temperature.

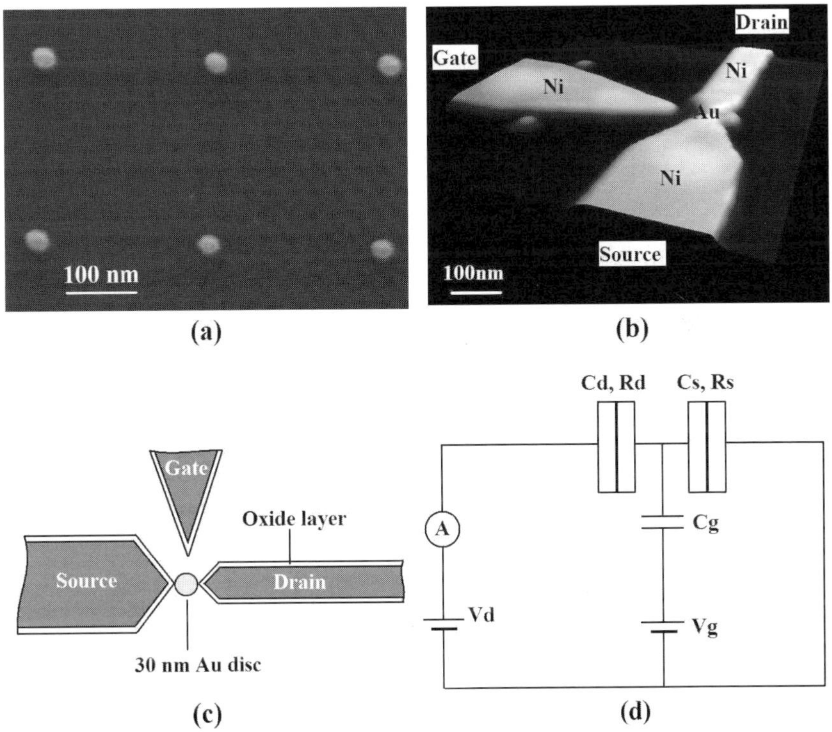

Figure 2.2 (a) SEM micrograph of the grid of Au discs. The Au disc is 30 nm in both diameter and thickness. (b) Atomic force micrograph of the ferromagnetic SET studied in the present work. The device is fabricated on top of a 100 nm thick SiO_2 layer. (c) Schematic picture of the F-SET studied in the present work. (d) Schematic circuit diagram for the F-SET.

Using the AFM manipulation technique described in Ref. 14, a gold nanodisc was positioned step-by-step with Ångström precision into the 25 nm gap between the O_2 plasma-processed drain and source electrodes. The resistance of the device was monitored in real-time while the Au disc was manipulated, and the desired resistance in the range

1 MΩ ~ 1 GΩ could normally be obtained after merely a few attempts. By re-positioning the Au disc in the gap it was also possible to tune the tunnel resistance. As previously mentioned, an important advantage of our device design compared to previous designs[14] is the presence of NiO tunnel barriers. In view of the open design of our device, with nanodiscs being manipulated into the gap between oxidized electrodes, we emphasize here the experimental evidence for the tunneling to occur through effective NiO barriers rather than through air barriers. For samples without O_2 plasma treatment following the Ar plasma etching, a device resistance of a few kΩ (or less) was observed after positioning the Au nanodisc in the gap. This stands in sharp contrast to the resistance of typically a few MΩ to GΩ mentioned above for the O_2 plasma treated devices. In addition, in O_2 plasma treated devices a re-positioning of the Au nanodisc in the gap resulted in small changes in the resistance in contrast to a dramatic change by several orders of magnitude observed after merely a few Ångström re-positioning in devices with air barriers. The distance from the gate electrode to the central Au disc was ~ 70 nm, which resulted in a much stronger coupling to the central island as compared to more conventional substrate back contacts. In Figs. 2.2b & 2.2c we show an AFM micrograph and a schematic picture of an assembled Ni/Au/Ni F-SET, respectively. A fairly high fabrication yield of typically 10% was obtained. The main factor limiting the yield stems from PMMA residues remaining on the sidewall of the front-ends of the electrodes. Also, occasionally, Au discs can be difficult to manipulate after the lift-off process. After fabrication, extensive transport measurements were performed at 4.2 K in a liquid helium Dewar (schematic circuit diagram is shown in Fig. 2.2d).

In order to investigate spin-transport phenomena in these fascinating F-SETs, it is obviously important to verify the magnetic switching behavior and spin-injection efficiency of the electrodes. The inset of Fig. 2.3 shows a SEM image of a Ni/NiO/Ni magnetic tunnel junction with overlapping Ni electrodes of the same dimensions as those used for the F-SET. The junction was fabricated by first forming a Ni electrode using conventional methods. Following this, a sequence of plasma treatment steps formed the NiO tunnel barrier on top of the electrode. Subsequently, a second Ni top electrode was defined to overlap the bottom electrode with approximately 50 nm using a high-precision alignment procedure.

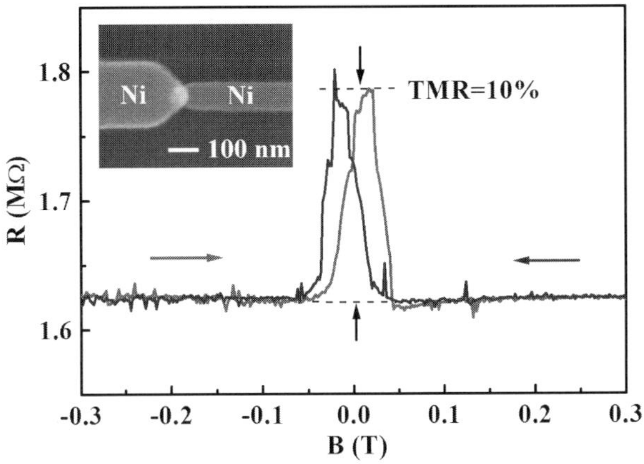

Figure 2.3 Switching behavior of a tunnel junction between two Ni electrodes separated by NiO as a function of magnetic field at 1.7 K. The inset shows a SEM micrograph of the device.

Magnetoresistance measurements clearly show a maximum TMR signal of about 10% while sweeping the magnetic field (Fig. 2.3). This TMR signal provides strong support for the premise that we can control the relative orientation of the magnetization of the two leads, and additionally, that the electrodes are efficient injectors of spin-polarized current. Using Julliere´s model[18], TMR = $(R_{AP}-R_P)/R_P = 2P^2/(1-P^2)$, where R_P and R_{AP} are the resistances in parallel and antiparallel magnetic configurations respectively, and P is the spin-polarization, we deduce a spin-polarization of 22% in the Ni electrodes. This value is in good agreement with the spin-polarization measured by Tedrow and Meservey in planar tunnel junction experiments[19].

2.3 Results and Discussions

Figures 2.4a and 2.4b show current-voltage characteristics at 4.2 K for a symmetric ($\Gamma_s = \Gamma_d$) and an asymmetric ($\Gamma_d >> \Gamma_s$) F-SET at on- (red curve) and off-state (blue curve), respectively, where Γ_s (Γ_d) refers to the electron tunneling rate between the source (drain) electrode and the central island. For the symmetric device, the tunnel rates are equal for the two tunnel junctions leading to Coulomb blockade only around zero drain-source bias, V_d.

It is noted that the current remains finite in the Coulomb blockade regime in the off-state for the symmetric device, which can be attributed to inelastic co-tunneling processes[20]. For the asymmetric device, $\Gamma_d >> \Gamma_s$ which results in a suppression of the current also at larger drain-source biases and an observed Coulomb staircase in the I-V curves. The Coulomb blockade is lifted when the island is charged with e/2 by applying a gate bias. Single electron tunneling through the device can thus be precisely controlled by tuning the capacitively coupled gate electrode.

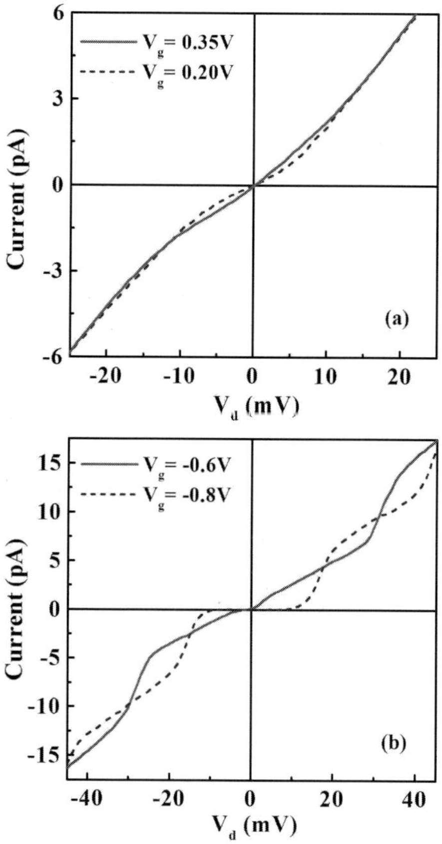

Figure 2.4 I-V characteristics at 4.2 K at different gate bias for (a) a symmetric (Γ_d=Γ_s) device and (b) an asymmetric ($\Gamma_d >> \Gamma_s$) device.

Figures 2.5a and 2.5b show color-coded plots of the differential conductance *dI/dV* as a function of V_d and gate voltage V_g (applied to the side-gate electrode) for the two devices. The dark areas correspond to low differential conductance and Coulomb blockade regimes for the symmetric and asymmetric devices, respectively. The differential conductance *dI/dV* is nonzero in the dark areas of Fig. 2.5a, as evident from the color-coded scale bar next to the figure, due to inelastic co-tunneling processes[20]. The shifts observed in the dark regions with respect to $V_g = 0$ in Fig. 2.5a (symmetric device) and Fig. 2.5b (asymmetric device) reflect the presence of a non-zero background charge.

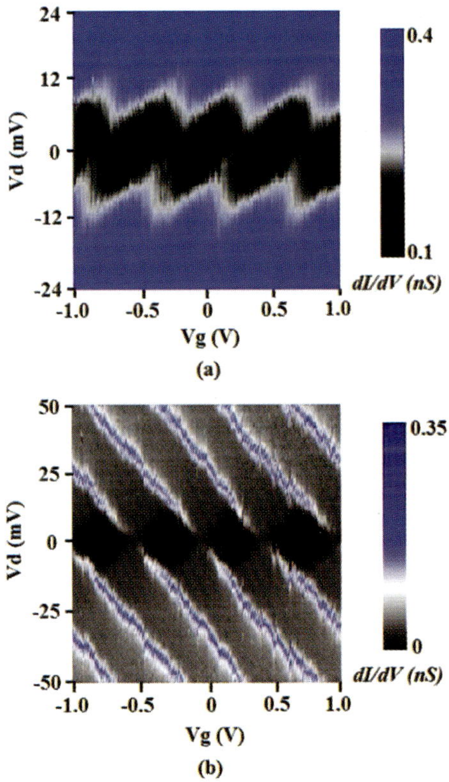

Figure 2.5 Color-coded plots of the differential conductance *dI/dV* as a function of drain-source bias V_d and gate voltage V_g for (a) the symmetric ($\Gamma_d = \Gamma_s$) device and (b) the asymmetric ($\Gamma_d >> \Gamma_s$) device. The measurements were performed at 4.2 K.

For the asymmetric device, the gate capacitance C_g, deduced from the spacing between neighboring degeneracy points at $V_d = 0$ where *dI/dV* is non-zero, is given by $C_g = e/(0.50\ V) = 0.32$ aF. The drain-source threshold voltage required for tunneling of one electron through the device is given by $V_{th} = e/C_\Sigma$ and amounts to about 10 mV. From this we estimate a total capacitance $C_\Sigma = C_s + C_d + C_g$ of 16 aF, and a corresponding charging energy $E_c = e^2/2C_\Sigma$ of 5 meV for the central island. The ratios C_s/C_g and C_d/C_g for the source and drain tunnel junctions, respectively, are determined from the slopes of the borders of the Coulomb blockade diamonds, and amounts to $C_s = C_d = 7.84$ aF. By performing numerical fittings of the experimental I-V curves following ref. 21, we extract tunnel resistances between the central island and the source and drain electrode, respectively, of $R_d = 150\ M\Omega$ and $R_s = 2.25\ G\Omega$. Due to the nonzero differential conductance in the Coulomb blockade region for the symmetric device, it is difficult to extract the degeneracy points at $V_d = 0$. To deduce the charging energy, we estimate a threshold voltage of about 4 mV in Fig. 2.4a from a linear extrapolation at large V_d to $I_d = 0$. From this we obtain a total capacitance $C_\Sigma = 40$ aF and a corresponding charging energy $E_c = 2$ meV. For simplicity, we assume $C_s = C_d \approx 20$ aF for the symmetric device and extract $R_s = 1.75\ G\Omega$ and $R_d = 1.6\ G\Omega$ from numerical fittings of the experimental I-V curves. In this analysis we have neglected co-tunneling processes. We point out here that the tunnel rates Γ_s and Γ_d depend not only on the capacitances and tunnel resistance but also on the effective voltage over the respective junction. In our case the bias was applied asymmetrically between the drain and source electrodes as shown in Fig. 2.2d. The high-conductance strips in Fig. 2.5b (asymmetric device) correspond to steps in the Coulomb staircase in Fig. 2.4b and indicate that the number of electrons on the island can also be precisely controlled with the drain-source bias in addition to the gate bias. Regarding the tunnel junctions as plate capacitors with $C = \varepsilon_r \varepsilon_0 A/r$, where A is the area and r is the thickness of the tunnel barrier, and using $C = C_s = 7.84$ aF for the asymmetric device, $\varepsilon_{NiO} = 10.31$ and r = 2 nm (estimated), we obtain an effective tunnel junction area of 174 nm^2 corresponding to 30 nm (height) × 5.8 nm (width) which is quite reasonable considering the geometric shape of the electrodes and the Au-island (as shown in Fig 2.2c). Due to the relatively large charging energy (2-5 meV), resulting from the dramatic decrease of the central island size, the single-electron transistor characteristics were present up to ~30 K in comparison to only a few tens - hundreds of mK observed in conventional ferromagnetic SET designs exhibiting a charging energy in the range 50~150 μeV[5-6, 11-12].

2.4 Conclusions

We have assembled robust high-quality ferromagnetic Ni/Au/Ni single-electron transistors by manipulating a single 30 nm Au disc to bridge the gap between the Ni source and drain electrodes. Tunnel junctions of NiO were formed by plasma oxidizing the electrodes prior to manipulation. The assembly scheme facilitates real-time tunable tunnel junction resistances. A side-gate fabricated close to the island allows a strong capacitive gate coupling. The small size of the central island facilitates typical single-electron transistor characteristics e.g. Coulomb blockade phenomena and periodic "Coulomb diamonds" at 4.2 K. In the asymmetric device, a distinct Coulomb staircase was observed. Typical charging energies of about 2-5 meV were deduced for the devices in this study. The unconventional assembly approach paves the way for future in-depth studies of the interplay between the Coulomb blockade effect, spin-dependent tunneling and spin relaxation mechanisms in nano-scaled F-SETs at elevated temperatures.

References

1. R. J. Schoelkopf, P. Wahlgren, A. A. Kozhevnikov, P. Delsing, and D. E. Prober, *Science* **280**, 1238 (1998).
2. M. H. Devoret and R. J. Schoelkopf, *Nature* **406**, 1039 (2000).
3. J. von Delft and D. C. Ralph, Phys. Rep. **345**, 61 (2001).
4. *Single Charge Tunneling*, NATO ASI Series, edited by H. Grabert and M. H. Devoret (Plenum, New York). Vol. **294**, 1 (1992).
5. K. Ono, H. Shimada, and Y. Ootuka, J. Phys. Soc. Jpn. **66**, 1261 (1997).
6. K. Ono, H. Shimada, and Y. Ootuka, J. Phys. Soc. Jpn. **67**, 2852 (1998).
7. *Spin Electronics,* edited by M. Ziese and M.J. Thornton (Springer, Berlin Heidelberg, 2001).
8. R.J. Elliot, Phys. Rev. **96**, 266 (1954).

9. Y.Yafet in *Solid State Physics*, edited by F. Seitz and D. Turnbull (Academic, New York), Vol. **14**, 1 (1963).
10. K. Yakushiji, F. Ernult, H. Imamura, K. Yamane, S. Mitani, K. Takanashi, S. Takahashi, S. Maekawa and H. Fujimori, *Nature Materials* **4**, 57 (2005).
11. J. H. Shyu, Y. D. Yao, C. D. Chen and S. F. Lee, J. Appl. Phys. **93**, 8421 (2003).
12. D. Wang and J. G. Lu, J. Appl. Phys. **97**, 10A708 (2005).
13. R. S. Liu, D. Suyatin, H. Pettersson and L. Samuelson, Nanotechnology **18**, 055302 (2007).
14. T. Junno, S.-B. Carlsson, H. Q. Xu, L. Montelius, and L. Samuelson, Appl. Phys. Lett. **72**, 548 (1998).
15. Y. Jaccard, Ph. Guttinne, D. Kelly, J.-E. Wegrowe, and J. -Ph. Ansermet, Phys. Rev. B **62**, 1141 (2000).
16. S.Y. Chou, P. R. Krauss, and L. Kong, J. Appl. Phys. **79**, 6101 (1996).
17. OOMMF is Object Oriented Micromagnetic Framework, a micromagnetic simulation code available free from NIST at http://math.nist.gov/oommf/.
18. M. Julliere, Phys. Lett. A **54**, 225 (1975).
19. R. Meservey and P.M. Tedrow, Phys. Rep. **238**, 173 (1994).
20. D.V. Averin and Y. V. Nazarov, Phys. Rev. Lett. **65**, 2446 (1990).
21. M. Amman, R. Wikins, E. Ben-Jacob, P. D. Maker, and R. C. Jaklevic, Phys. Rev. B **43**, 1146 (1991).

3

NANOPOROUS ALUMINA TEMPLATES FOR NANOWIRE ELECTRON DEVICES

Travis L. Wade, Costel S. Cojocaru, Jean-Eric Wegrowe and Didier Pribat

Nanoporous alumina templates are demonstrated as device structures for nanowire field-effect transistors (FETs). A ZnO nanowire surrounding gate FET demonstrates the utility of this nanostructured template. This bottom-up approach is unique in that the fabrication can be almost entirely done using room temperature electrochemistry. The technique is extended to lateral-anodized nanoporous aluminum thin-films, compatible with standard silicon technology and illustrated by the possible fabrication of an active matrix backplane with only four lithographic masks.

3.1 Introduction

There are many chemical and physical routes to nanowires and nanodots. Electrodeposition, sol-gel synthesis, and CVD are but a few of the techniques that can produce nanostructures with nanometer control[1-5]. The quality and reproducibility of these nanostructures are very well established. The real difficulty, however, is to organize and contact these nano-objects[6, 7]. This is when nanoporous templates become very useful.

The role of the template is two fold: First, it allows the production of the structure with the best possible reproducibility and it plays the role of a skeleton in order to organize the different functions and building blocks of a device, the active components and the different interfaces (active elements, electric contacts, various bias voltages, etc.) on a rigid body. The role of the template is to allow the manipulation of the nanoscale building blocks without the help of a microscopic tip (no systematic use of AFM or STM) or without the need of top-down processes (lithography, FIB, lift-off). Second, this nanoscaffold is used to link the structure to the macroscopic world, i.e. the contacts without the use of (nano) lithography. Also, template structuring/growth brings an easy solution to the problem of end-to-end registration [8].

In the scheme of synthesis inside the templates, it is possible to identify three different steps: (1) the creation of the building blocks, *e.g.*, the nanowires or nanodots, (2) the assembly of the nano building blocks into a functional architecture within the template and (3) the fabrication and control of the contacts to the macroscopic world. The first and second steps coincide for metallic nanowires or semiconductors that are made by electrodeposition. For carbon nanotubes and silicon nanowires, a catalytic layer is made by electrodeposition followed by CVD for the carbon nanotubes or vapor-liquid-solid growth for the silicon nanowires. The final step, and perhaps the most important and difficult, is the contact of the nanoscale objects to the macroscopic world.

3.2 Basic Background

Anodization of a metal is the controlled anodic growth of a metal-oxide film on a metal surface, mainly aluminum, in an electrolytic bath[9-18]. There are two types of metal-oxide films formed by anodization, amorphous barrier films and porous films. These oxide films can be a micron thick for barrier films to many tens of microns thick for porous films, as opposed to the 2-3 nanometer thick metal-oxide films that exist on many metals as a result of ambient atmospheric oxidation. The barrier layer films are formed in pH neutral aqueous electrolytes, such as ammonium borate, in which aluminum is insoluble. Porous films are formed in acidic aqueous electrolytes such as dilute, 1 molar, sulfuric acid, in which the oxide layer is formed but also dissolves at the same time, enhanced by the local electric field.

For these oxide films, the metal to be anodized is the positive electrode, anode, in an aqueous electrolyte solution with an inert negative electrode, cathode, to complete the circuit. When a potential of several volts is applied to the cell, current flows from the anode to the cathode, the electrode/solution interfaces are polarized, and electrochemical reactions occur at these interfaces. The important reactions for us occur at the anode (Figure 3.1a). Here a metal oxide film is grown due to the field-induced migration of cations, Al^{3+}, from the electrode and anions, O^{2-}, from the solution. The cations react with water at the oxide/solution interface, equation 1, and the anions react with the metal/oxide interface, equation 2. The oxide growing at the metal/oxide interface is pure oxide and the oxide growing at the oxide/solution interface incorporates anions from the electrolyte. The reaction occurring at the cathode results in the production of hydrogen, equation 3. The overall reaction is thus oxide film growth and hydrogen gas formation, equation 4.

$$2Al^{3+} + 3H_2O \rightarrow Al_2O_3 + 6H^+ \quad (1)$$

$$2Al + 3O^{2-} \rightarrow Al_2O_3 + 6e^- \quad (2)$$

$$6H^+ + 6e^- \rightarrow 3H_2 \quad (3)$$

$$2Al + 3H_2O \rightarrow Al_2O_3 + 3H_2 \quad (4)$$

In the case of neutral electrolytes, i.e. Al^{3+} insoluble, the barrier layer grows to less than 1 micron with applied voltages of 500-700 V. After this thickness is obtained, the film undergoes dielectric breakdown. Anodization in an acidic electrolyte such as sulphuric acid, however, changes the structure of the oxide layer, Figure 3.1b. This is caused by two phenomena: First, the Al^{3+} cations are now soluble in the solution so the oxide layer dissolves at the same time it is forming. The thickness of the oxide layer is proportional to the voltage and inversely proportional to the solubility of the aluminium in the solution or the competition between oxide growth and dissolution; the more acidic the solution the more soluble the aluminium and the thinner the oxide layer and the smaller the pores. Second, the mechanical stress between the aluminium and the oxide layer volume expansion of 1.2 [14] results in heterogeneous dissolution of the oxide layer or in the formation of pores. The dissolution of the aluminium is field assisted so the oxide dissolves in areas where the oxide layer is thinner. The pores become deeper and other areas become isolated from dissolution. The competition between oxide dissolution and growth modulated by the film stress can result in an ordered porous structure, in stabilizing regimes of the Laplace pressure of the pores and the elastic stress[19,20], Figure 3.1b. The heterogeneous nature of the oxide film is the same as the case of the barrier film with the pure oxide is light blue, Figures 1a and b, and the anion contaminated oxide is light green, Figures 3.1a and 3.1b, while the pores are depicted as black, Figure 3.1b. The dimensions and interpore spacing of the pores are proportional to the anodization voltage[12]. Also, as stated earlier, the more soluble the aluminum the smaller the diameter of the pores, 5-30 nm for sulphuric acid, 40-60 nm for oxalic acid, and 80-130 nm for phosphoric acid, Table 3.1. These values coincide with the increasing pH's of the solutions. For all electrolytes the oxide layer is thinnest at the pore bottom where it is dissolving. The pore diameter, Φ_p, and the interpore distance, D_{int}, are proportional to the anodisation voltage U[1]. In other words, $\Phi_p = k_1U$ and $D_{int} = k_2U$, where $k_1 \sim 1.29$ nm/V and $k_2 \sim 2.5$ nm/V[14, 21].

Electrodeposition of materials in porous structures is performed by connecting the aluminum part of the template to the working electrode lead of a potentiostat. This sample is then placed in a metal salt electrolyte such as $ZnNO_3$. Potentiostatic electrodeposition of the material (M) is then performed in the pores by means of the working electrode, where M (Co, Cu, Ni, Zn etc.) is a metal and n is the number of electrons, equation 5.

$$M^{n+} + ne^- \rightarrow M^0 \qquad (5)$$

During the electrodeposition the metal ions diffuse into the pores to the bottoms and are deposited by applying a potential between the working and the reference electrode, which is kept at a constant value for potentiostatic control. The counter anode serves as the current source for the system. A tremendous variety of materials and structures can be deposited by electrodeposition: homogeneous metals, semiconductors, and heterogeneous structures[1].

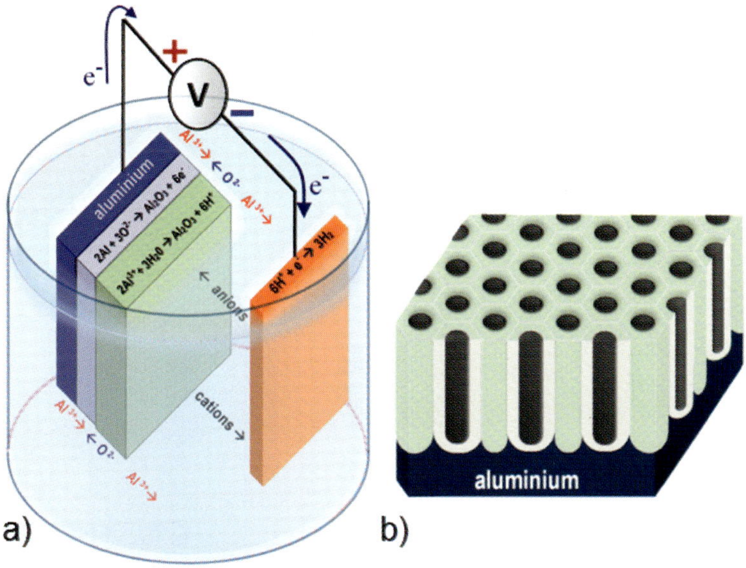

Figure 3.1 a) Reactions during the anodic formation of alumina. b) Schematic of nanoporous anodized aluminum.

Chemical vapor deposition (CVD) can be used in order to synthesize semiconductor nanowires inside the pores of alumina membranes. The preferred way of growing such nanowires is by using the so-called vapour-solid-liquid (VLS) method which was developed some 40 years ago by Wagner[22,23]. In the VLS method, the growth of semiconductor materials is mediated by eutectic mixtures which are in the liquid state at the operating temperature. Semiconductor atoms (*e.g.*, Si) originating from the pyrolysis of a gas precursor (*e.g.*, SiH_4) are incorporated into the liquid eutectic alloy (*e.g.*, Au-Si, with 19 atom % Si). Upon saturation, the liquid starts to precipitate the excess semiconductor (*e.g.*, Si). A permanent regime is rapidly reached, whereby the flux of semiconductor (Si) atoms reaching the liquid-solid interface equals the flux of semiconductor atoms (Si) incorporated in the liquid eutectic at the vapour-liquid interface. Growth is highly anisotropic because the sticking coefficient of gas phase molecules on the liquid surface is much higher than that on surrounding solid surfaces, including the side walls of the growing wire/whisker. Figures 3.2a-c summarise the situation for the case of Si wire synthesis.

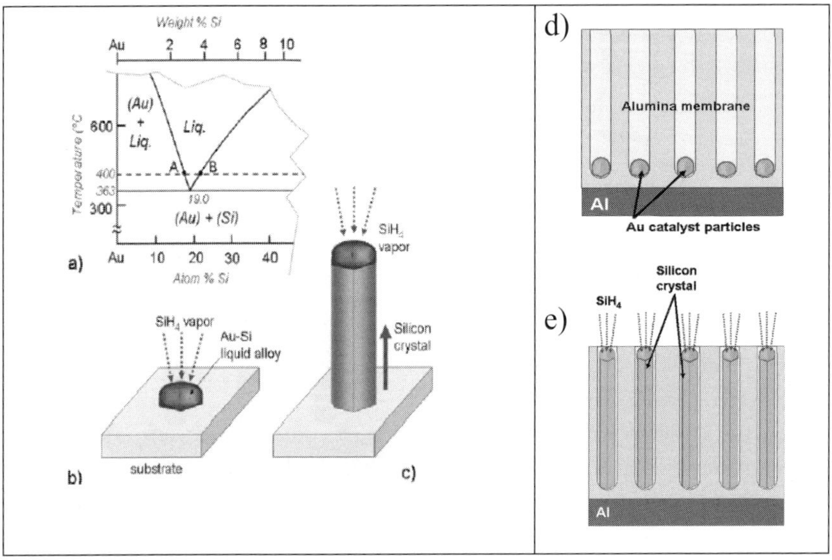

Figure 3.2 Schematic mechanism of the VLS growth process for Si and its implementation in a porous alumina membrane: a) Au-Si phase diagram (on the Au-rich side), b) formation of the Au-Si eutectic alloy from an Au cluster/dot, c) growth of a Si whisker, d) alumina membrane with electrodeposited gold particles at the bottom of the pores, and e) VLS growth of Si nanowires inside the alumina membrane.

The VLS process has recently been applied to the growth of Si nanowires (NWs) and high performance field effect transistors (made with individual NWs) have been demonstrated, exhibiting mobility values up to ~ 1300 cm^2/Vs[24-26]. However, the problem is to organise these NWs on the surface of a substrate, in order to control their placement and in-plane organisation. For this purpose, fluidic methods based on the use of Langmuir-Blodgett films have been developed[27], but they do not provide end-to-end registration. Template growth provides a way to organise an ensemble of nano-objects, providing precise placement, control, and registration.

When template growth of semiconductor NWs inside anodic alumina membranes is of concern, the first task is to deposit metal particles at the bottom of the pores (Figure 3.2d). Subsequent alloying with semiconductor atoms (released by decomposition of precursor molecules, e.g., SiH_4) will induce the formation of the eutectic alloy which is necessary for initiating the VLS growth process. Figure 3.2e schematically shows the situation.

The feasibility of the VLS growth of Si NWs inside the pores of anodic alumina has already been demonstrated for vertical templates[28, 29]. Figure 3.3 shows Si NWs grown in such a way. In order to ease the observation, the growth time has been purposely extended, resulting in NWs growing out of the membrane (Figure 3.3a). Transmission electron microscope (TEM) analysis shows that the NWs are monocrystalline. As already observed for NWs grown in "free space" (i.e. not constrained by the template structure) the crystallographic orientation seems to depend on the diameter[30].

3.3 Novel Approaches

The template growth situation depicted in Figures 3.2e and d (with NWs perpendicular to the surface of the substrate) is well adapted to two-terminal devices. The substrate can be used as a first contact electrode and the second electrode can be deposited on top of the membrane after growth of the NWs[7]. However, for three-terminal devices, there is a topological problem, since it is complicated (although not impossible) to incorporate a third electrode between the substrate and the top electrode[7,32-34]. In order to circumvent this problem, new types of membrane geometries in which the pores develop from the source contact and in the same plane as the source[35].

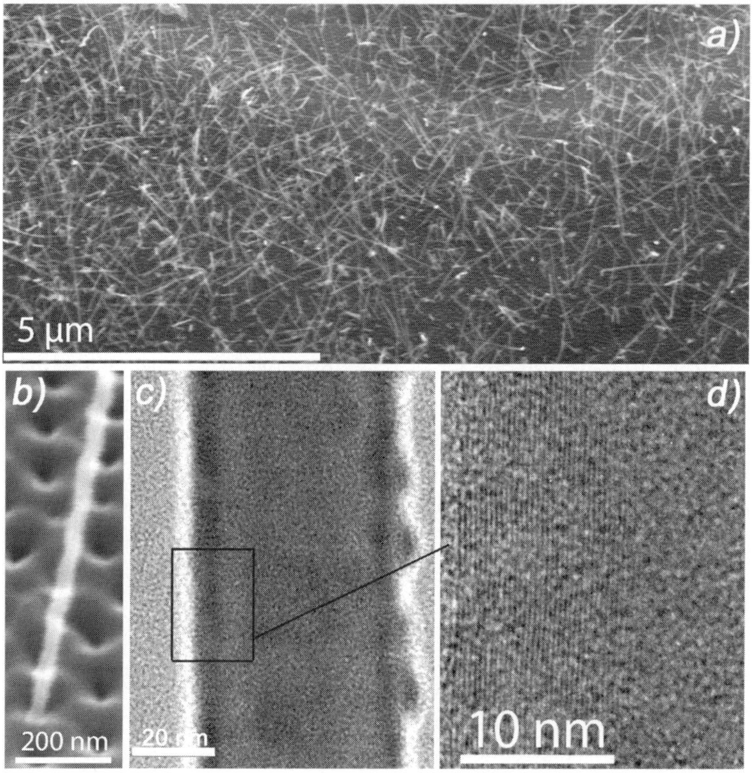

Figure 3.3 a) Top view of Si nanowires grown in a vertical anodic alumina membrane. For easier observation, the growth time has been prolonged purposely, which results in nanowires growing out of the membrane. b) Close view of a nanowire emerging from a pore c), and d) transmission electron microscope views of a nanowire. Note the amorphous layer at the nanowire surface on d). Crystallographic planes can be observed on d).[31]

We have developed two novel processes for the synthesis of alumina templates that allow placement of a third electrode close enough to the nanowires to induce an electric-field effect[7,36]. The first is shown in Figure 3.4. The gray colored area of the diagram is an aluminum wire, the end of which has been electrochemically etched to about 3 microns or less in diameter, the bottom 10 mm is anodized about 500 nm deep perpendicular to its axis to form an isolating layer, green. The bottom

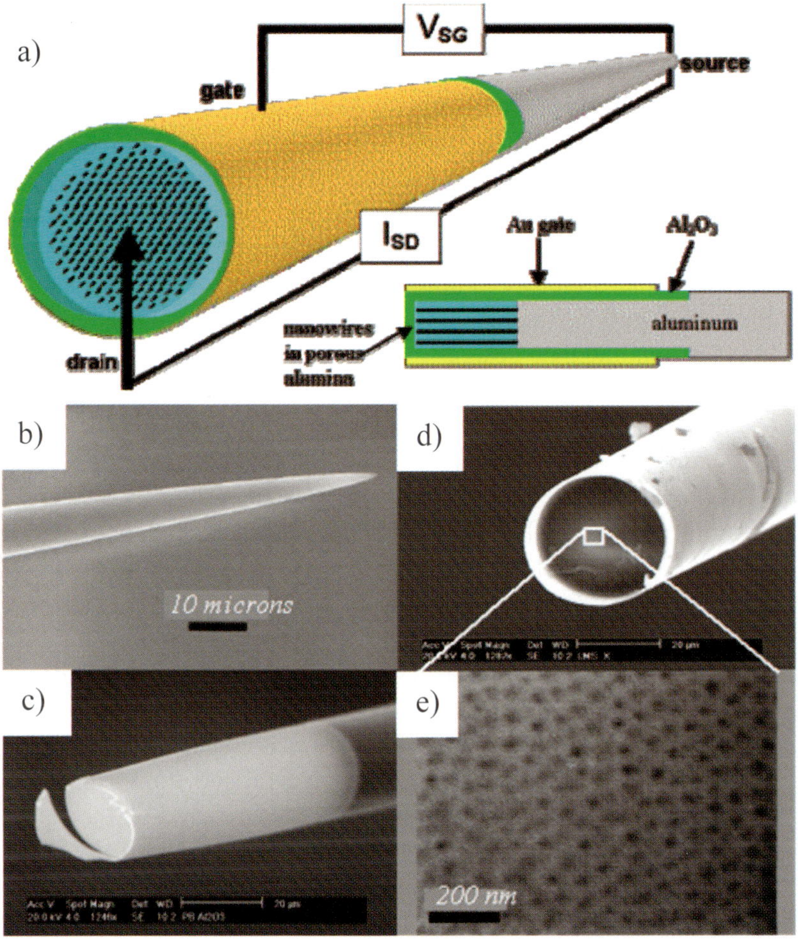

Figure 3.4 a) Scheme of a bi-directionally anodized wire transistor. The gray area is the aluminum wire and the yellow area is the gold gate separated from the aluminum wire by the external oxide, green. The nanowires are the black dots in the light blue internal porous alumina template or the black lines in the cutaway scheme on the lower right. b-e) SEM micrographs of some of the steps for template fabrication: Aluminum wire is electrochemically etched to a tip diameter of a few microns b). It is then anodized perpendicular to its axis to form an insulating layer onto which a gate electrode is sputtered. Next, the wire is cut and the interior is etched, c). It is then anodized in the interior to form a network of pores parallel to the wire axis as a template for the synthesis of nanowires d) and e).

5 millimeters of the isolating layer is sputter coated by gold as a gate electrode, yellow. The sputtered gold layer does not cover all of the oxidized area and is thus isolated from the aluminum. A very small section of the bottom is cut off to expose the interior aluminum. This is then anodized parallel to the wire to form a porous alumina template about one micron deep, sky blue. The nanowires, black, are electrodeposited into the parallel template by connecting the aluminum wire as the cathode of an electrochemical cell. The nanowires are connected as the drain, the aluminum wire as the source, and the gold layer as the gate for transport measurements. This is essentially a vertical surrounding gate design[37-41].

The SEM photos in Figure 3.4 illustrate various steps involved in the fabrication of an aluminum nanowire transistor. Aluminum, 120 micron diameter, wire is the starting material for the transistor template. This is annealed at 500°C for 24 hours in a vacuum tube furnace and cut into 3 cm long pieces. The wire was then electrochemically etched in a 25% $HClO_4$, 75% ethanol solution at +10 V to a few microns in diameter, Figure 3.4b. The etch rate is approximately 1.5 microns per second. Next, the wire is anodized at +40 V in 0.3 M oxalic acid for 2-4 minutes to produce an external isolating layer on the exterior of the wire. At this voltage the growth rate of the oxide layer about 150-200 nanometers per minute. The tip of the wire is immersed in a 1 M NaOH solution to dissolve the isolating oxide layer at the tip. This is then electrochemically etched. Now, a gold layer is sputtered on the exterior, insulating layer of the wire. The gold layer, which is electrically isolated from the interior aluminum, will function as a gate electrode. The wires from Figures 3.4d and e are anodized internally at +40 V in a 0.3 M oxalic acid for 5 minutes to form an internal nanoporous template. The diameter of the wire in this photo is about 30 microns, which is at least ten to thirty times larger than is used for the transistor. The larger diameter was necessary to observe the interior anodization, Figure 3.4e. Anodization on very small areas has been observed[35, 42]. The pores are 40-50 nanometers in diameter and should be about one micron deep. This will serve as the template for the electrodeposition of the nanowires. The nanowires can then be grown and contacted with the gate electrode already in place.

The idea of anodizing Al wires and tubes has been tried for applications such as extracting fibers or chromatography columns[43-45], however, using anodized wires as templates and multidirectional anodization is a new concept. These templates could be used for transistors, microelectrodes, diodes, micro/nano fluidics, and moulds for MEMs etc.

ZnO and other transparent conducting oxides are interesting as materials for UV lasers, light-emitting diodes, photo detectors, and for applications in flat panel displays and solar cells. ZnO has a band gap of 3.35 eV and is normally a n-type direct gap material[46-49]. When ZnO is doped with transition metals it can form spin-polarized light sources[47].

Once the porous template is made, ZnO nanowires can be electrodeposited into it. This is done by potentiostatic electrodeposition of ZnO from a 0.001M $Zn(NO_3)_2$ solution (pH~ 6.8) at –1.500 V vs. a saturated calomel electrode (SCE) in the pores for 1000 seconds. Since this is an unbuffered solution the polarization causes an increase of the pH at the sample surface, due to a loss of H^+ by H_2 gas formation, which results in the precipitation of ZnO[50, 51]. Optimization of the solution and an increase in the solution deposition temperature could produce single crystal ZnO nanowires[51,52]. When the ZnO arrives at the surface and extends beyond the template it can be electrically contacted. Then the aluminum wire base of the template and the gold gate electrode can be easily contacted. Chemical analysis by backscattering of the electrodeposited ZnO on Au substrates revealed stoichiometric ZnO, however, the XRD analysis showed many phases of polycrystalline ZnO and Zn metal.

Figure 3.5a shows the room temperature drain current vs. the drain-source voltage (I-V) for different gate potentials[53]. The very low currents could be due to the high resistance of ZnO nanowires as has been seen in other studies[54]. The threshold voltage depends on the source-drain potential[53]. The channel dimensions should be the same as that of the nanowire, 1 micron long and 40 nanometers in diameter. No effort was made to remove the oxide barrier layer at the pore bottoms after the interior anodization, although there is a contact which could result in a tunneling barrier between the ZnO nanowires and the aluminum source and contribute the high resistance of the device. The off current of 20 pA is seen for all samples and may possibly be a background current in the measurement instruments or a tunneling current. The gate-source capacitance of this structure is rather large since the gate covers five millimetres of the device. This aspect of the device was not optimised and the gate area could easily be reduced by masking. The signal was stable for about 20 cycles but slowly declined, possibly due to depletion of charge carriers by irreversible electromigration or screening of the gate. Figure 3.5b shows the transfer characteristics at a drain-source bias of 1.0 V and shows that the transistor operates as a p-channel depletion mode device. From 3.5b the on/off ratio is only two.

The second new approach, lateral nanoporous alumina membranes, allows integration of alumina templates into standard silicon technology. Aluminum thin-film stripes are sandwiched between two insulating layers (Figure 3.6a) and locally etched to yield the structure schematically shown on Figure 3.6b. Individual Al stripes are electrically contacted, away from the locally etched area, and partially immersed in an electrochemical bath for anodic oxidation, Figure 3.6c. Because of the engineered structure of Figure 3.6b, the electric field (during anodic oxidation) is forced to develop parallel to the surface of the substrate. Hence the pores in the anodic alumina are also forced to develop parallel to the surface of the substrate as schematically shown on Figure 3.6d.

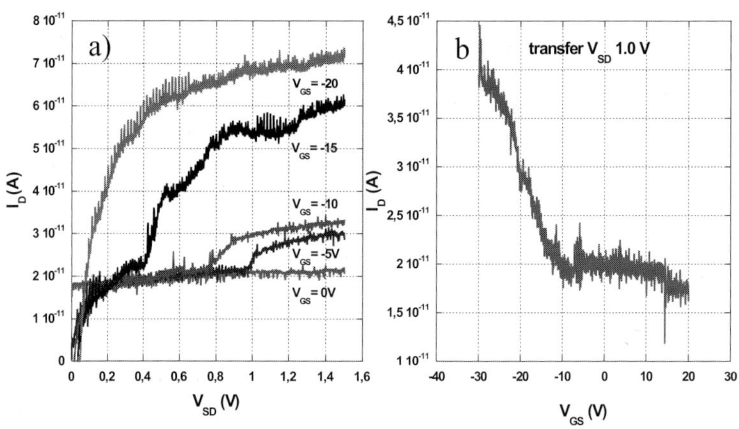

Figure 3.5 a) Drain current vs. drain-source potential at different gate potentials. b) The transfer characteristics of the ZnO transistor at a drain-source potential of 1.0 V.

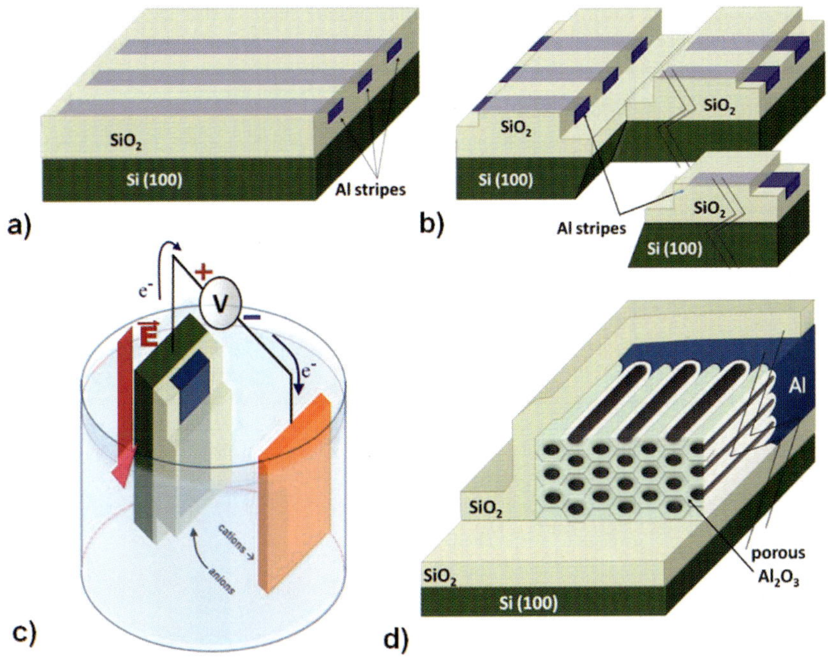

Figure 3.6 Synthesis of lateral nanoporous alumina membranes.

We have carefully studied the process of lateral anodic oxidation, depending on the various operating parameters[55]. Table 3.1 shows the variation of the pore diameter with respect to the anodization voltage and type of electrolyte. As explained above, concerning vertical pores, large pore diameters are formed by using phosphoric acid, intermediate diameters are formed by using oxalic acid and small pore diameters are obtained by using sulfuric acid. The various results are summarized in Table 3.1 below. Note that the pore diameter can be enlarged by chemical etching in phosphoric acid. Using a H_3PO_4 concentration of 5% wt, the etch rate of anodic alumina is ~ 8 nm/h at 20°C.

Table 3.1. Conditions for lateral anodization in various acidic solutions (H_2SO_4, $H_2C_2O_4$, H_3PO_4). The pore diameter scales between ~ 5 and 100 nm depending on conditions.

Acid	Anodization conditions			SEM image
	T-°C	V-V	t-sec	Φ –Pore diameter
Sulphuric 0.5 M	0	3	300	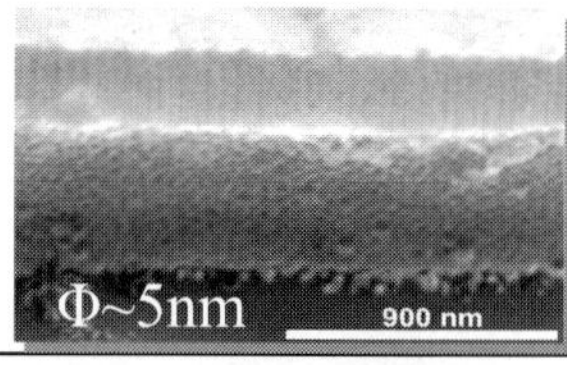
Oxalic 0.5 M	0	20	300	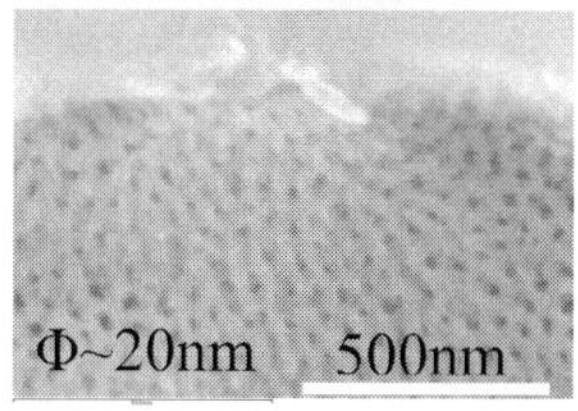
Oxalic 0.5 M	0	25	300	
Oxalic 0.5 M	0	40	300	

Table 1 *Continued*

Oxalic 0.5 M	0	50	3000	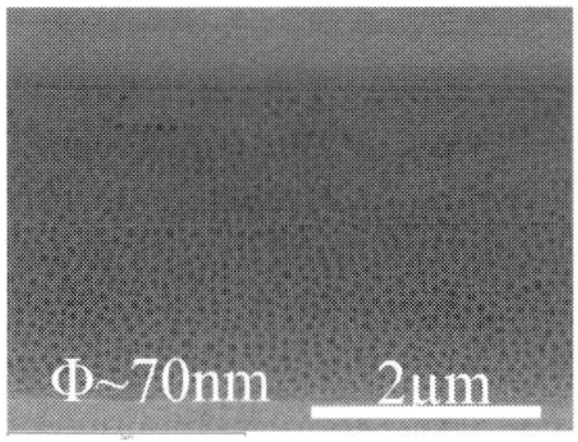 After 20 min Pore widening in 0.5M H_3PO_4
Phosphoric 0.5 M	0	100	3000	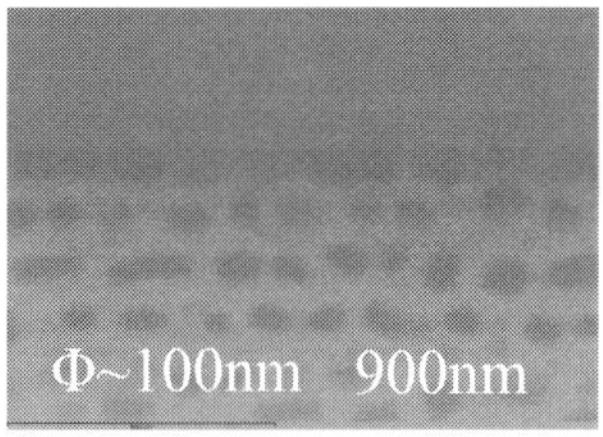

At this point, we would like to emphasize that the filling of lateral alumina templates with nanowires should be a transposition of what has already been demonstrated for vertical templates. The interest of the lateral templates is illustrated and explained below (see Figure 3.7) for the fabrication of an active matrix liquid crystal display backplane with a reduced number of masks[56].

An Al thin film (data lines) is first deposited and etched (mask # 1), so as to yield a "finger" in each pixel (Figures 3.7a and 3.7d, the later being a top view). The etched Al film is then capped with a deposited insulating layer (e.g., SiO_2, Figure 3.7b). This capping insulating layer is etched (mask # 2) at the tips of the Al fingers in each pixel (Figures 3.7b and d). Anodic oxidation of the Al film is then performed locally, at the end wall of the tip of the Al fingers, resulting in the structure shown on Figure 3.7c. Gold nanoparticles are then electrodeposited at the bottom of the pores of the formerly synthesised porous anodic alumina fingers (Figure 3.7e, cross section).

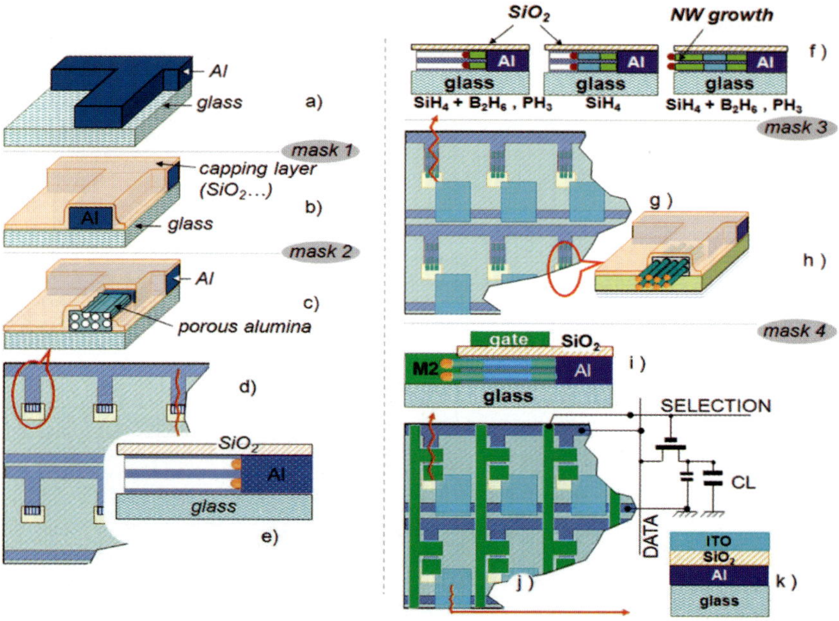

Figure 3.7 Proposed process steps for the fabrication of the active plate of an active matrix liquid crystal display based on the use of lateral porous anodic alumina followed by VLS growth of Si nanowires.

After completion of the lateral templates and Au electrodeposition, growth of the Si NWs is performed inside the pores of the lateral templates at the tip of each Al finger (Figure 3.7f), using sequentially a mixture of SiH_4 + dopant, pure SiH_4 and again a mixture of SiH_4 + dopant. This can be done in one pump-down operation and results in the structure shown on Figure 3.7f. A transparent oxide film (e.g., indium-tin-oxide, ITO) is then deposited and etched (mask # 3, see Figure 3.7g, top view), in order to define the pixel area. Finally, the gate metallization (select lines) is deposited and etched (mask # 4, Figure 3.7j). Note that the gate metal also connects the drain of the TFT to the transparent electrode of the pixel. For the sake of clarity, Figure 3.7i and k show cross-sectional views of the TFT and storage capacitor. We would like to emphasize that only 4 masks are used for the fabrication of the AMLCD backplane and that the thin-film transistor structure is fabricated during only one "pump-down" operation, as with amorphous silicon (a-Si:H) TFTs (see ref. 56).

3.4 Conclusion

Nanoporous alumina shows great potential as a template material for hosting various nanowires. The ability to easily control the pore diameter and length is a tremendous advantage, as well as the fact that alumina is high-temperature resistant. This means that a wide variety of materials can be synthesized in alumina-based nanoporous templates, from room temperature growth of metallic nanowires and semiconductors, such as ZnO, to high temperature growth of carbon nanotubes and silicon or germanium nanowires. Alumina templates also provide a structure that organizes the nanowires for device fabrication and contacting. As an example, microscope observations of silicon nanowires grown in these templates showed the feasibility of synthesizing high quality materials. Aluminum wires were electrochemically sculptured into bi-directional templates for the growth and contacting of nanowires as three terminal devices. The gate dependent drain currents and transfer characteristics demonstrate that it is possible to make nanowire transistor devices in a beaker without clean rooms or lithography. In-plane anodized aluminum thin films were shown as templates compatible with standard silicon technology. As with vertical porous structures, their versatility is clearly seen by the variety of pore dimensions that can be obtained. Finally, we described a detailed scheme to incorporate nanoporous alumina structures into an active matrix backplane, thus reducing the mask count to only 4.

References

1. C. N. R. Rao; F. L. Deepak; G. Gundiah; A. Govindaraj 2003 *Progress in Solid State Chemistry* **31** 5-147
2. B. Doudin; A. Blondel; J. P. Ansermet 1996 *Journal of Applied Physics* **79** 6090-6094

3. S. Dubois; J. L. Duvall; A. Fert; J. M. George; J. L. Maurice; L. Piraux 1997 *Journal of Applied Physics* **81** 4569-4569
4. W. Schwarzacher; O. I. Kasyutich; P. R. Evans; M. G. Darbyshire; G. Yi; V. M. Fedosyuk; F. Rousseaux; E. Cambril; D. Decanini 1999 *Journal of Magnetism and Magnetic Materials* **199** 185-190
5. P. Levy; A. G. Leyva; H. E. Troiani; R. D. Sanchez 2003 *Applied Physics Letters* **83** 5247-5249
6. L. Piraux; A. Encinas; L. Vila; S. Màtéfi-Tempfli; M. Màtéfi-Tempfli; M. Darques; F. Elhoussine; S. Michotte 2005 *Journal of Nanoscience and Nanotechnology* **5** 372-389
7. T. L. Wade; J.-E. Wegrowe 2005 The European Physical Journal Applied Physics **29** 3-22
8. S. Jin; D. M. Whang; M. C. McAlpine; R. S. Friedman; Y. Wu; C. M. Lieber 2004 *Nano Letters* **4** 915-919
9. J. P. O'Sullivan; G. C. Wood 1970 Proceedings of the Royal Society of London Series a-Mathematical and Physical Sciences **317** 511-&
10. G. E. Thompson; R. C. Furneaux; G. C. Wood; J. A. Richardson; J. S. Goode 1978 *Nature* **272** 433-435
11. H. Masuda; K. Fukuda 1995 *Science* **268** 1466-1468
12. O. Jessensky; F. Muller; U. Gosele 1998 *Applied Physics Letters* **72** 1173-1175
13. A. P. Li; F. Muller; A. Birner; K. Nielsch; U. Gosele 1998 *Journal of Applied Physics* **84** 6023-6026
14. K. Nielsch; J. Choi; K. Schwirn; R. B. Wehrspohn; U. Gosele 2002 *Nano Letters* **2** 677-680
15. K. Nielsch; F. Muller; A. P. Li; U. Gosele 2000 *Advanced Materials* **12** 582-586
16. K. Nielsch; R. Hertel; R. B. Wehrspohn; J. Barthel; J. Kirschner; U. Gosele; S. F. Fischer; H. Kronmuller 2002 *IEEE Transactions on Magnetics* **38** 2571-2573
17. Y. T. Pang; G. W. Meng; L. D. Zhang; W. J. Shan; X. Y. Gao; A. W. Zhao; Y. Q. Mao 2002 *Journal of Physics-Condensed Matter* **14** 11729-11736
18. Y. H. Wang; Y. Q. Xu; W. L. Cai; J. M. Mo 2002 *Acta Physico-Chimica Sinica* **18** 943-946

19. G. K. Singh; A. A. Golovin; I. S. Aranson; V. M. Vinokur 2005 *EUROPHYSICS LETTERS* **70** 836-842
20. G. K. Singh; A. A. Golovin; I. S. Aranson 2006 *Physical Review B* **73** 205422
21. D. Crouse; Y. H. Lo; A. E. Miller; M. Crouse 2000 *Applied Physics Letters* **76** 49-51
22. R. S. Wagner; C. J. Doherty 1966 *Journal of the Electrochemical Scoiety* **113** 1300-&
23. R. S. Wagner; W. C. Ellis 1964 *Applied Physics Letters* **4** 89-&
24. Y. Cui; Z. H. Zhong; D. L. Wang; W. U. Wang; C. M. Lieber 2003 *Nano Letters* **3** 149-152
25. X. F. Duan; N. C. M.; V. Sahi; J. Chen; J. W. Parce; S. Empedocles; J. L. Goldman 2003 *NATURE* **425** 274-278
26. J. Goldberger; A. I. Hochbaum; R. Fan; P. D. Yang 2006 *Nano Letters* **6** 973-977
27. P. D. Yang 2003 *NATURE* **425** 243-244
28. K. K. Lew; J. M. Redwing 2003 *Journal of Crystal Growth* **254** 14-22
29. D. Pribat; C. S. Cojocaru; A. Rumayantseva 2005 Proceedings of the 1st International TFT Conference (ITC 2005) p. 160
30. Y. Wu; Y. Cui; L. Huynh; C. J. Barrelet; D. C. Bell; C. M. Lieber 2004 *Nano Letters* **4** 433-436
31. D. Pribat; C. S. Cojocaru; M. Gowthan; L. Eude; P. Bondavalli; P. Legagneux 2007 *Journal of the Society for Information Displays* **in press**
32. X. Hoffer; C. Klinke; J.-M. Bonard; L. Gravier; J.-E. Wegrowe 2004 *EUROPHYSICS LETTERS* **67** 103-109
33. J.-E. Wegrowe; S. E. Gilbert; D. Kelly; B. Doudin; J.-P. Ansermet 1998 *IEEE Transactions on Magnetics* **34** 903-905
34. J. F. Dayen; T. L. Wade; M. Konczykowski; J. E. Wegrowe; X. Hoffer 2005 *Physical Review B* **27** 073402
35. C. S. Cojocaru; J. M. Padovani; T. Wade; C. Mandoli; G. Jaskierowicz; J.-E. Wegrowe; A. F.-i. Morral; D. Pribat 2005 *Nano Letters* **5** 675-680
36. T. Wade; J.-E. Wegrowe, Procede de fabrication de composants electroniques et composants electroniques obtenus par ce procede. Demande de brevet d'invention 2868201, date de dépôt 23.03.2004.

37. T. Bryllert; L. E. Wernersson; L. E. Froberg; L. Samuelson 2006 *IEEE Electron Device Letters* **27** 323-325

38. J. Goldberger; A. I. Hochbaum; R. Fan; P. Yang 2006 *Nano Letters* **6** 973-977

39. T. Schulz; W. Rösner; E. Landgraf; L. Risch; U. Langmann 2002 *Solic-State Electronics* **46** 985-989

40. J. Chen; R. Konenkamp 2003 *Applied Physics Letters* **82** 4782-4784

41. A. K. Sharma; S. H. Zaidi; S. Lucero; S. R. J. Brueck; N. E. Islam 2004 *IEE Proc.-Circuits Devices Syst.* **151** 422-430

42. H. Zhang; C. Zhi; T. Li; K. Saito 2005 *Journal of Nanoscience and Nanotechnology* **5** 1745-1748

43. D. Djozan; Y. Assadi; S. H. Haddadi 2001 *Analytical Chemistry* **73** 4054-4058

44. D. Djozan; A.-Z. M. 2003 *Surface & Coatings Technology* **173** 185-191

45. W.-S. Chae; S.-J. Im; J.-K. Lee; Y.-R. Kim 2005 *Bulletin of the Korean Chemical Society* **26** 409-412

46. Z. Fan; D. Wang; P.-C. Chang; W.-Y. Tseng; J. G. Lu 2004 *Applied Physics Letters* **85** 5923

47. Y. W. Heo; D. P. Norton; L. C. Tien; Y. Kwon; B. S. Kang; F. Ren; S. J. Pearton 2004 *Materials Science and Engineering R* **47** 1-47

48. Y. W. Heo; L. C. Tien; Y. Kwon; D. P. Norton; S. J. Pearton 2004 *Applied Physics Letters* **85** 2274-2276

49. H. T. Ng; J. Han; T. Yamada; P. Nguyen; Y. P. Chen; M. Meyyappan 2004 *Nano Letters* **4** 1247-1252

50. M. Pourbaix, *Atlas d'Equilibres Èlectrochimiques*. Gauthier-Villars: Paris, 1963; Vol. Chapter 15 Zinc.

51. S. Peulon; D. Lincot 1998 *Journal of the Electrochemical Society* **145** 864-874

52. Y. Leprince-Wang; G. Y. Wang; X. Z. Zhang; D. P. Yu 2006 *Journal of Crystal growth* **287** 89-93

53. T. L. Wade; X. Hoffer; A.-D. Mohammed; J.-F. Dayen; D. Pribat; J.-E. Wegrowe 2007 *Nanotechnology* **18** 125201

54. O. Harnack; C. Pacholski; H. Weller; A. Yasuda; J. M. Wessels 2003 *Nano Letters* **3** 1097-1101

55. M. Gowtham, L. Eude, C.S. Cojocaru, B. Marquardt, H.J. Jeong, P. Legagneux, K.K. Song and D. Pribat, accepted for publication in *Nanotechnology*
56. In the AMLCD industry, it is critical to reduce the number of masking steps in order to decrease the cost of fabrication of displays. *See e.g., D. Pribat, Materials Science Forum, Trans Tech Publications, Switzerland, vols. 455-456, 2004, p. 56*

4

SINGLE-WALLED CARBON NANOTUBE TRANSISTORS

Sunkook Kim and Saeed Mohammadi

We describe approaches taken to address the challenges in making useful and reproducible single-walled carbon nanotube field effect transistors. Both Al_2O_3 and HfO_2 dielectrics for gate insulation have been discussed as are techniques for aligning carbon nanotubes. A variety of measurements, including transistor output characteristics, transfer characteristics and low frequency noise measurements have been described to assess the properties of these devices.

4.1 Introduction

Since the discovery of carbon nanotubes in 1991[1], carbon nanotubes (CNTs) have been the focus of extensive research for many potential applications including sensing, chemistry, biology, and electronic devices. CNTs are rolled up sheets of graphite with diameter of 1.2 nm and larger and are available in both single-walled and multi-walled forms. Because of their perfect one-dimensional crystalline structure, CNTs exhibit unusual physical, chemical, mechanical and electrical properties.

Attempts to make first CNT field effect transistors started in 1998[2]. The filed has since exploded by theoretical and experimental researches that have looked into potential advantages and applications of such devices. There have thus been many publications in this field over the past few years. As suggested by these articles, single-walled carbon nanotube field-effect-transistors (SWNT-FETs) not only present a simple implementation for high performance nanoscale devices, but also are very attractive devices due their following outstanding electronic properties:

1) Ballistic transport
Due to their one dimensional structure, backscattering of carriers is strongly suppressed in SWNTs. It leads to ballistic transport, meaning that the mean free path for charge carriers is longer than the channel length. In the literature[3], carrier mobility as high as 8000 $cm^2V^{-1}sec^{-1}$ is reported, which is already superior to the mobility of Si transistors with mobility values of 1400 $cm^2V^{-1}sec^{-1}$. Ballistic transport of SWNT-FETs can yield very high speed transistors, necessary for high performance microprocessors and memory modules.

2) Flexible Device
The strong covalent bonding of C-C affords the CNTs high mechanical and thermal stability. Furthermore, CNTs are transparent under visible light. Softness and transparency of CNTs can be utilized in thin film transistors for display devices.

3) Low power dissipation
An IBM group has compared the electrical properties of SWNT-FETs to both conventional high performance p-type Si-transistors and SOI MOSFETs[5] as illustrated in Table 4.1. Important differences are in the off current I_{off}, transconductance, and sub-threshold swing. For VLSI circuits, the subthreshold swing S, defined by $S = \log[dV_{gs}/d(\log(I_{ds}))]$, is a key parameter for a transistor to switch current. Clearly, a lower sub-threshold slope is more attractive since it is desirable to have the transistor "turn off" as close to the threshold voltage as possible. SWNT-FETs feature very low subthreshold slopes despite the fact that their fabrication

technology is still in infancy. Additionally, smaller off current I_{off} in SWNT-FETs means lower static power dissipation.

Table 4.1: Comparison of SWNT-FETs to Si-MOSFETs and SOI MOSFETs.

	SWNT-FET[4]	P-MOSFET	P-SOI MOSFET
Gate length (nm)	150	15	50
Gate oxide thickness (nm)	8	1.4	1.5
Ion (μA/μm)	33	265	650
Ioff (nA/μm)	0.05	<500	9
Subthreshold swing (mV/dec)	70	~ 100	70
Transconductance (μS/μm)	3000	975	650

4) Low cost
Instead of PNP or NPN structures of regular MOSFETs and MESFETs, SWNT-FETs are composed of a metal-SWNTs-metal structure. Only a few metallization steps are sufficient to build up transistors without elaborate processing such as implantation, post annealing, and etching.

Regardless of the high performance of SWNT-FETs, integration of individual SWNTs has been impeded by uncontrolled structural variations in SWNTs. During the synthesis of carbon nanotubes, variations in the chirality (arrangement of the carbon atom on the surface of the nanotube) and diameter result in either metallic or semiconducting carbon nanotubes. These properties cause major problems in the realization of large area and / or complex circuits based on carbon nanotubes, as listed in the following.

(1) Difficulty in the separation of semiconducting and metallic single-walled carbon nanotubes.
(2) Alignment of the nanotubes to the device structure in a controlled and reproducible fashion.
(3) Devising a fabrication process that yields devices isolated from their environment.

In this chapter, we will focus on approaches taken to address the above challenges, necessary to develop future SWNT-FET fabrication techniques. First, the electronic structure of SWNTs and the operation principles of SWNT-FETs are briefly presented. Second, high-k dielectrics (Al_2O_3 and HfO_2) for the realization of high performance SWNT-FETs are introduced. Then, techniques to fabricate both room temperature devices based on dielectrophoresis and high temperature devices based on CVD synthesis of nanotubes on quartz substrates are illustrated. Finally, intrinsic properties of the interface between SWNT and its gate oxide are investigated, based on direct current as well as low frequency electronic noise measurements.

4.2 Electronic Structure and Operating Principles of SWNT-FETs

Single-walled carbon nanotubes (SWNTs) are rolled up sheets of graphene with diameters of 1.2 nm and larger. The roll-up direction of graphene determines the chirality of nanotubes. Metallic nanotubes have armchair chirality while semiconducting nanotubes are arranged with zig-zag chirality. In semiconducting SWNTs, the energy band gap[6] is inversely proportional to the diameter of the carbon nanotube according to:

$$E_{GAP} = \frac{4\hbar v_F}{3d} \approx \frac{0.8}{d}(eV) \qquad (1)$$

where d is the diameter of SWNT and v_F is the Fermi velocity. The diameter of a typical SWNT is in the range of 0.5 nm to 3 nm. Therefore, small diameter SWNTs have a wide band-gap of 1.6 eV, but large diameter SWNTs show narrow band-gaps, as low as 0.25 eV.

Recent SWNT-FETs[4,7] with Pd metal contacts operate as Schottky barrier transistors. Carrier transport and switching in SWNT-FETs are controlled by the Schottky barrier between source metal and SWNTs. IBM researchers have demonstrated that the Fermi energy level moves up toward conduction band by removing oxygen from the environment[5]. Generally, intrinsic SWNT-FETs in air form a P-type transistor, where a positive gate voltage turns the device off as shown in Fig. 4.1(a). In P-type transistors, Pd contacts provide the lowest barrier between metal and nanotube which results in the lowest contact resistance. However, in case of vacuum (i.e. absence of oxygen), the characteristic of the transistor changes from p-type to n-type as the Fermi energy moves into the conduction band. In our experiments, SWNTs grown on Si wafer show p-type characteristic but SWNTs grown on quartz substrates show an ambipolar characteristic (both N-type and P-type) as shown in Fig. 4.1(b). To explain the ambipolar SWNT-FETs, we assume that the Fermi-energy (E_F) is located around the middle of the band gap (E_g). For positive gate bias, the valence band moves up and the Schottky barrier gets narrower, resulting in an increase in the hole tunneling current. With the application of a negative gate bias, the conduction band moves down and the Schottky barrier becomes narrower, resulting in an increase in the electron tunneling current. The ambipolar characteristics may become useful in implementing CMOS logic for potential digital circuit applications. To suppress ambipolar characteristics the following post process steps are suggested by various research groups: (1) Field manipulation[8], (2) Potassium doping[9], (3) Polymer gating[10] and (4) Adsorption of O_2 gas[4].

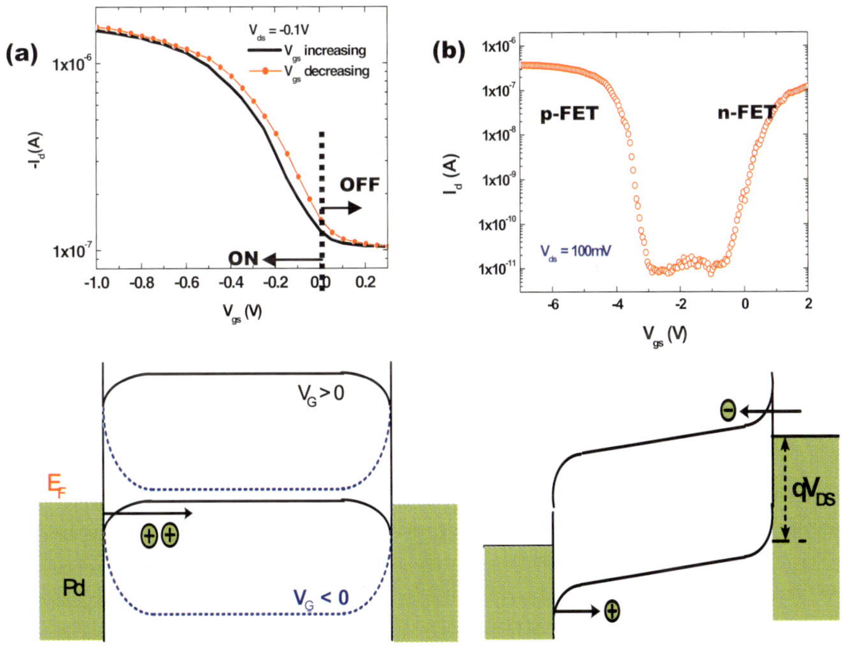

Figure 4.1: (a) The I_d-V_{gs} curve and energy band diagram of SWNT-Pd metal for p-type transistor (b) The I_d-V_{gs} curve and energy band diagram of SWNT-Pd or Al metal for ambipolar transistor.

4.2 High-k Dielectrics (Al_2O_3, HfO_2) for Advanced SWNT-FETs

Significant progress has been made in improving the electrical performance of SWNT-FET devices. Current research is focused on achieving high transconductance through scaling down the gate dielectric and utilizing novel gate dielectric materials. Several groups have investigated new SWNT-FET gate dielectric structures including (i) bottom-gate structure (gate underneath the nanotube) with native Al_2O_3 dielectric[11], (ii) top-gated with 15-20 nm thick SiO_2 dielectric[12] and (iii) top-gated with Atomic Layer Deposited (ALD) 8-20 nm thick high-k zirconium oxide (ZrO_2)[13] and hafnium oxide (HfO_2)[14] dielectrics. Among the above techniques, utilizing high-k ALD ZrO_2 and HfO_2 as gate dielectric is very promising and has resulted in transistors with near-ballistic transport and close to ideal sub-threshold swing of ~ 60 mV/ decade.

In the following section, we will study high performance and high interface quality top-gated SWNT-FETs with Al_2O_3 and HfO_2 dielectric films deposited by the ALD technique.

4.2.1 Atomic layer-deposited high-k dielectrics

Among the various techniques for growing gate dielectrics on CNTs, the ALD method is of particular interest because high-k gate films based on ALD provide high capacitance, resulting in effective carrier injection into the channel and high device transconductance. ALD is an ultra-thin-film deposition technique based on sequences of self-limiting surface reactions enabling thickness control on atomic scale even for complicated 3-D structures. The ALD process could also be performed at a low temperature (25-300°C), suitable for processes with CNTs. High-quality dielectric film and good interface property (*i.e.* low interface trap density) in high-k / CNT system can be achieved through the optimization of ALD growth conditions and Post Deposition Annealing (PDA). In our current experiments, the optimized conditions for ALD oxide deposition were as follows.

1) ALD Al_2O_3 film: Al_2O_3 films were grown on CNTs using alternating pulses of $Al(CH_3)_3$ (the Al precursor) and H_2O (the oxygen precursor) at 300°C in a carrier (N_2) gas flow. Post deposition annealing was done in N_2 ambient at 600°C for 30 seconds, using Rapid Thermal Processing (RTP). Al_2O_3 has a large bandgap of ~ 9 eV, a high

breakdown electrical field of 5 to 30 MV/cm, high permittivity of 8.6-10 and high thermal stability up to at least 1000°C.

2) ALD HfO_2 film: High-k HfO_2 films were deposited using an ASM Microchemistry F-120 ALCVD™ Reactor. 20 nm ALD HfO_2 layers were grown on SWNTs at 300°C by using $HfCl_4$ and H_2O as precursors. During deposition, High-k HfO_2 films were crystallized, thus the PDA process was not required.

4.2.2 Fabrication and performance of SWNT-FETs

Utilizing Thermal Chemical Vapor Deposition (Thermal CVD) and Plasma Enhanced Chemical Vapor Deposition (PECVD) to synthesis CNTs are two of the common techniques to fabricate SWNT-FETs. SWNT-FET devices shown in the inset of Fig. 4.2 were fabricated on a high resistivity Si substrate ($\rho \approx 1$ MΩ) with a 500 nm thermal silicon dioxide layer. Single-walled carbon nanotubes (SWNTs) were synthesized on the thermal oxide by thermal CVD of methane on the substrate using commercial ferritin (Sigma) catalyst. The grown carbon nanotubes had a diameter of 1 to 3 nm with a uniform tube density of 2-10/μm^2. Following the nanotube growth, a 15 nm thick ALD aluminum oxide (Al_2O_3) film was deposited on SWNTs using the ASM

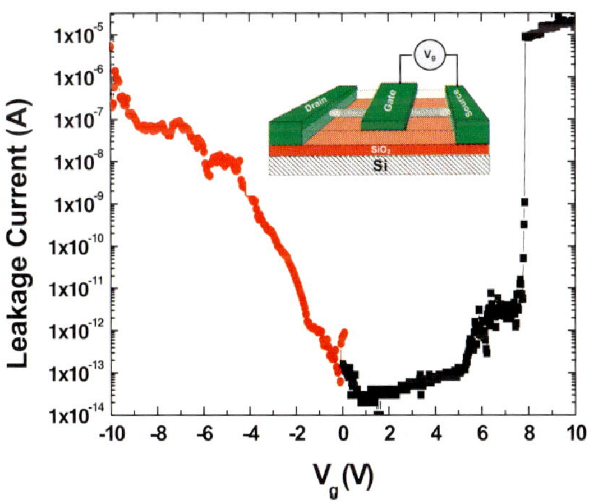

Figure 4.2: Leakage current (A) versus gate bias V_g (V) for SWNT-FETs having thin ALD aluminum oxide. (Inset) Schematic diagram of a top-gate single-walled carbon nanotube field effect transistor with ALD oxide as the gate dielectric.

Microchemistry F-120 ALCVD^{TM} Reactor. Using wet etching (diluted HF solution), the ALD oxide on the source and drain patterns was removed while the gate area was protected by the photoresist. Metal contacts were formed by electron beam deposition of Pd followed by a lift-off process. Gate metal was defined by UV photolithography followed by the deposition of Ti/Au (10/50 nm) with minimum gate length of 1.5 μm. Ti/Au (20/450 nm) metal interconnects were finally deposited on top of the source and drain Pd contacts.

Figure 4.2 shows the gate leakage current of a SWNT-FET having 1.5 μm gate length with a semiconducting nanotube in the channel. As can be seen from the figure, the leakage current is very small, in the range of 10 fA – 10 pA for gate voltage in the range of $-2.5\text{ V} < V_g < 7\text{V}$. The leakage current is about five orders of magnitude smaller than the nanotube drain current.

Figure 4.3 shows the SEM top view of the transistor and I-V characteristics of the same transistor with various gate dielectric thicknesses. In Figs. 4.3 (b) and (c) I-V characteristics for a 15 nm thick Al_2O_3 is shown while Figs. 4.3 (d) and (e) show the results for a transistor with a 20 nm thick HfO_2 gate dielectric. HfO_2 has a higher dielectric constant (κ ~ 20-30) than Al_2O_3 (κ ~ 8.6-10) resulting in higher drain current capability, higher transconductance and smaller sub-threshold swing of transistors fabricated with HfO_2 despite the slightly thicker dielectric layer used (20 nm for HfO_2 vs. 15 nm for Al_2O_3).

Table 4.2: Electrical performance of SWNT-FET with ALD Al_2O_3 and HfO_2

	SWNT-FET with Al_2O_3	SWNT-FET with HfO_2
Gate oxide thickness (nm)	15	20
Gate length (μm)	1.5	1.5
I_{on} (μA)	11	13
I_{off} (nA)	100	10
Subthreshold swing (mV/dec)	105	100
Transconductance (μS/μm)	1500	4000

Figures 4.3(b) and (d) show typical I_d versus V_{gs} characteristics when V_{gs} is swept forward and backward. In SWNT-FETs, the direction of V_{gs} sweep and its speed creates a hysteresis response due to the interface charge traps, mainly from water molecules[9]. The amount of

voltage shift depends on the interface trap density on / near SWNTs and how fast the sweep is performed. The small hysteresis observed in Fig. 4.3 indicates that ALD Al_2O_3 and HfO_2 suppress traps in SWNT-FETs. Figure 4.3(c) and (d) show I_d versus V_{ds} characteristics at different gate biases. The curves resemble a conventional P-MOS, exhibiting a linear region at low $|V_{ds}|$ and a saturation region at higher $|V_{ds}|$. Device performances of both SWNT-FETs are summarized in Table 4.2.

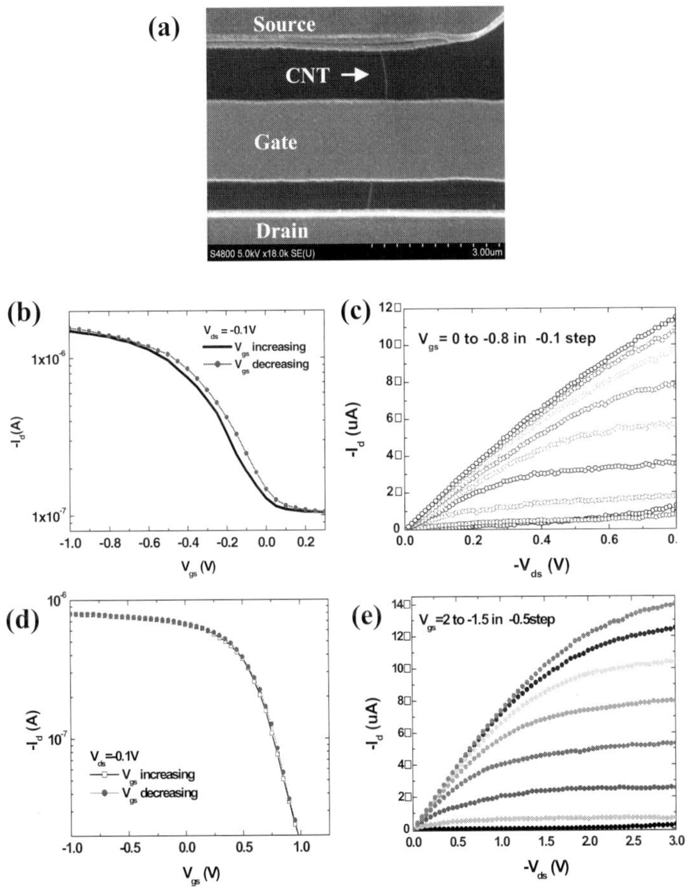

Figure 4.3: Characteristics of a p-type SWNT-FET with 15 nm thick ALD Al_2O_3 (Figs. (b) and (c)) and 20 nm thick ALD HfO_2 (Figs. (d) and (e)) as gate dielectrics. (a) SEM image of SWNT-FET (b), (d) Current I_d vs V_{gs} as increasing (black) and decreasing gating sweeps (red) for SWNT-FET at V_{ds} of -0.1V. (c), (e) Drain current vs drain bias as a function of gate bias for the same device.

4.3 Alignment Techniques for Carbon Nanotubes

A lack of large-scale integration techniques has been a major obstacle in the development of SWNT-FETs for practical circuit applications. One of the major challenges in achieving such integration is to align and position individual nanotubes. Up to now, various methods for selective deposition or growth of SWNTs on two electrodes have been developed[15-16]. In this section, we discuss two different alignment techniques that have been devised. The first technique is a room temperature device fabrication technology based on dielectrophoresis. The second technique is a high temperature device fabrication method based on CVD synthesis of nanotubes on quartz wafers. In the dielectrophoresis technique, individual SWNTs suspended in ethanol display a positive dielectrophoresis characteristic, which helps in achieving highly spatial deposition. Common issues are bundling of nanotubes due to their covalent affinity and large contact resistances due to the large tunneling resistance among bundled SWNTs. We have performed electrical measurements of transistors fabricated using dielectrophoresis and measured transistors with high contact resistance of 2.3 MΩ, low mobility, and no clear saturation regime.

On the other hand, in the development of aligned arrays of SWNTs on quartz, parallel SWNTs arrays with a controllable number and spacing are utilized to construct the SWNT-FETs. This technology opens a promising route to position SWNTs in a specific area for practical implementation of SWNT-FETs. This should allow the device designer to minimize contact capacitance and contact resistance through device geometry optimization. Transistors using aligned arrays of SWNTs achieve high mobility, low contact resistance and high on-current with clear saturation behavior at high drain source voltages.

4.3.1 SWNT-FET fabrication using dielectrophoresis

In this technology we focus on the self-assembly technique for separating, orienting and positioning of bundles of single-walled carbon nanotubes (SWNTs) on electrodes separated by a few microns. By applying an alternating electric field, a single bundle of SWNTs is selectively deposited between two small electrodes. The fabrication of CNT transistors based on this approach occurs at room temperature. Electrical burning can help eliminating unwanted metallic nanotubes that are otherwise present in the bundle to achieve higher fabrication yield.

In our experiments, SWNTs with a diameter of ~ 1.5 nm synthesized by laser ablation from Carbon Nanotechnologies Inc were used. Source and drain metal contacts were formed by photo-lithography and lift-off processing. The metal patterns were deposited on a thermally grown SiO_2 layer with a thickness of t_{ox} = 150 nm grown on a P+ Si substrate. The gap between the electrodes was 3 or 4 μm, and the width of the electrodes was 5 μm. The single-walled nanotubes suspended in ethanol were diluted to the extent that the liquid appeared transparent. (concentration = 10 ng/ml). Ultrasonic vibration for several hours was used to untangle the nanotubes to small bundles and single nanotubes.

Dipole of a SWNTs bundle should be induced under a non-uniform external electric field. Due to the interaction between the dipole and the external field, dielectrophoresis acting on the SWNT bundles was generated. A combination of AC and DC electric fields was applied. The dielectrophoresis was induced by the AC electric field, while the DC electric field produced a mechanical flow. The frequency of AC electrical field was kept constant at 5 MHz. Once the two electric fields were established, diluted ethanol with suspended nanotubes was applied on the SiO_2 surface. The composite electric field was applied until the ethanol solution dried completely at room temperature. Following the ethanol evaporation, Pd was deposited and patterned to form the contacts on the nanotubes.

In order to find the optimal deposition and alignment condition, the amplitude of combined DC and AC signals was changed step by step. Figure 4.4 shows SEM images of single carbon nanotube bundles trapped between two electrodes. The deposition of nanotube bundles is mainly dependent on the mechanical flow generated by the amplitude of the DC electric field. The optimal condition of AC signal for the deposition and alignment is an electric field of magnitude 0.6 V/μm at a frequency of 5 MHz. The experiment was highly reproducible with average bundle diameter of about 15 nm.

The extracted transconductance of the device $g_m = dI_d/dV_{gs}|_{Vds=1V}$ was found to be about 50 nS for devices with one nanotube bundle. The maximum current showed several hundreds of nA range. Current saturation was not observed due to the high contact resistance of 2.3 MΩ, in spite of the fact that Pd was used to lower the contact resistance. The

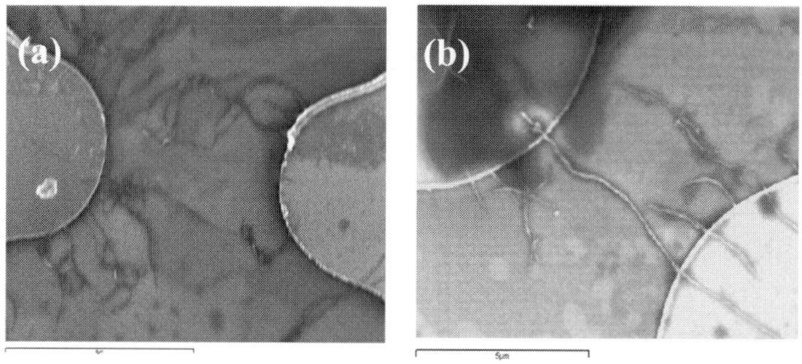

Figure 4.4: SEM images of two different aligned patterns from two different alignment conditions. (left) condition: DC = 0.7 V, AC = 5 V_{PP}, Frequency = 5 MHz (failed) (right) condition: DC = 0 V AC = 2.3 V_{PP}, Frequency = 5 MHz (well-aligned).

poor performance is attributed to the weak coupling between adjacent SWNTs in a bundle that leads to a high tunneling resistance of 2 - 140 MΩ[17].

4.3.2 SWNT-FETs using guided growth of nanotube arrays on quartz

In this technology, aligned arrays of SWNT are grown by thermal CVD on miscut single-crystal quartz substrates. This method is used to improve the mobility of aligned arrays of SWNT-FETs compared to what can be achieved from random network devices[18]. Aligned arrays of SWNTs can sustain the intrinsic mobility of individual SWNTs due to the minimum interaction of SWNTs. Prior to fabrication, a ST-cut quartz substrate obtained from Hoffman Materials Inc. was annealed for 8 hours at 900°C in air. The annealing resulted in the generation of nanoscale grooves on the quartz wafer. Nearly perfect alignment of SWNTs could be achieved with direct growth of nanotubes inside these grooves, as illustrated in Fig. 4.5. The degree of alignment was influenced by the annealing of quartz before tube growth[16]. Increasing the annealing time, which may increase the degree of order in the crystal lattice near the surfaces as well as the lengths and order of the steps, improved the alignment of grown SWNTs. Figure 4.5(c) shows the SEM image of a corner of the patterned catalyst. SWNTs are aligned parallel to the ST-miscut quartz in the middle of the two catalysts, while tube-tube crossing is shown near the catalyst patterns. In our experiments, the source and drain were patterned as 3 μm long islands in the middle of two catalyst patterns with a spacing of 20 μm. The arrays of SWNTs between source and drain were perfectly parallel, as shown in Fig. 4.5(d) and Fig. 4.5(e). Figs. 4.5(f) and (g) show the IV characteristics for a 1.5 μm gate length SWNT-FET with a 50 nm thick Al_2O_3 layer utilizing aligned arrays of SWNTs. The maximum intrinsic transconductance of 0.5 μS and the

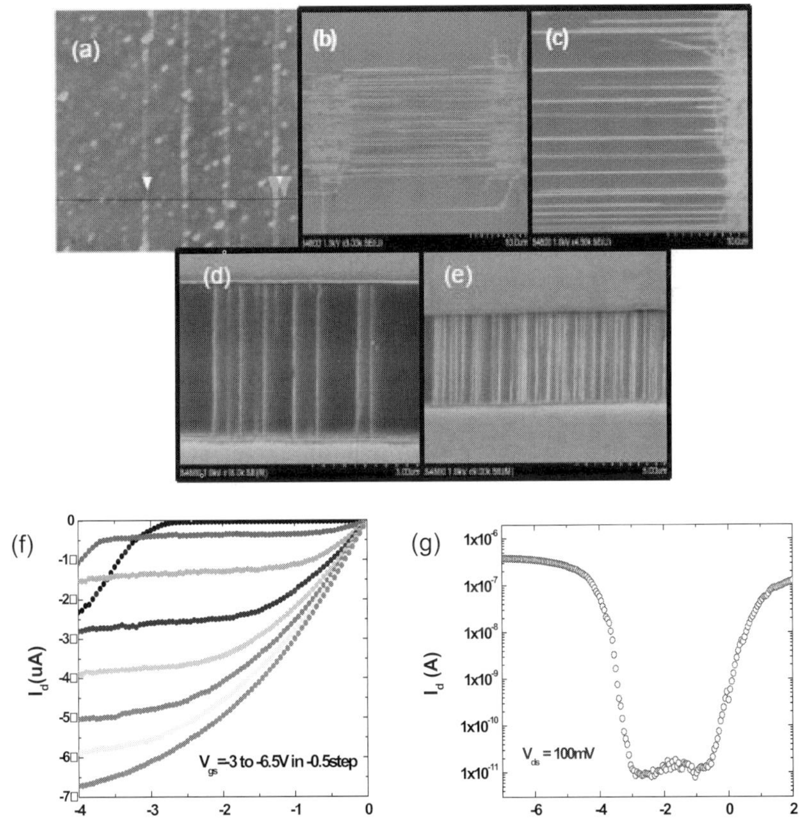

Figure 4.5: (a) AFM image of aligned SWNTs on ST-miscut quartz substrate. The grown SWNTs have diameters of 0.8-1.7 nm and lengths of 10-30 μm with a spacing of 200 nm on quartz wafer. (b) SEM images of grown aligned arrays of SWNTs using patterned iron catalysts with a spacing of 20μm. (c) The enlarged image of the edge of the catalyst patterns. (d) and (e) SEM images of perfectly aligned arrays of SWNTs between source and drain electrodes with spacing of 3 μm and widths of 6 μm and 15 μm. (f) I_d-V_{ds} characteristics of an aligned array of SWNT-FETs with 50 nm thick Al_2O_3 oxide. (g) Transfer characteristics of the same device.

maximum on-current of 7 μA were observed, which is at least one order of magnitude larger than the one achieved for SWNT-FETs based on dielectrophoresis. The on / off ratio with a 6 μm source-width SWNT-FET was 4 x 10^4, where three SWNTs were connected between the source and the drain. The IV curves of SWCNT-FETs showed an ambipolar behavior, exhibiting an operating regime at positive and negative V_{gs} as shown in Fig. 4.5(g). Unlike the bundled transistors fabricated by dielectrophoresis, current saturation was clearly observed in these devices. The estimated drain and source metallization contact resistance was around 38 kΩ. In this case, Pd metallization was efficient in lowering the contact resistance, since individual nanotubes could be covered by Pd metallization.

4.4 Interface Quality of SWNT-FETs

Despite advancements in placing and aligning nanotubes, there are still two important obstacles in achieving high performance electronic devices using carbon nanotubes. These are high access resistance of nanotube / contact metallization due to the Schottky nature of the contact and interface quality of gate-oxide / nanotube interface. In this section, characterization of this interface quality is presented. The complications of low quality interface are large amplitude of low frequency noise as well as hysteresis in the IV characteristics of SWCNT-FETs. Nanotube-based devices reported in the literature have exhibited high low frequency noise characteristics[19-21]. High amplitude of low-frequency noise increases the minimum detectable signal at low frequencies and can have an adverse effect on non-linear circuits implemented using SWCNT through noise up-conversion.

Low frequency noise characterization as well as hysteresis in V_g-I_d characteristics can be used to study the interface quality of SWNTs prior to and after ALD oxide deposition. ALD Al_2O_3 and HfO_2 gate oxides serve as a passivation layer for SWNT to minimize undesirable current drift from the nanotube-ambient environment. Previous work points to hysteresis in CNT-FETs as a method of analyzing the interface quality qualitatively. Generally, in conventional metal-oxide-semiconductor (CMOS) technology, capacitance-voltage (C-V) measurement and low frequency noise are used to analyze the interface traps in oxide / semiconductor structures. The carbon nanotube interface has an extremely small gate dielectric capacitance, $C_{ox} \sim 2\pi\varepsilon_0\varepsilon / \ln(2t_{ox}/R) \sim 28$ *aF/nm*, much lower than the minimum detectable value of current laboratory equipment. Therefore, C-V measurement cannot be used to measure the interface quality of SWNT-FETs. Low frequency noise is then the only quantitative method for the analysis of interface traps in CNT devices. It has been found that the 1/f noise in passivated SWNT-FETs has a significant dependence on the device transconductance[22]. In this section, we also propose a simple model for low frequency noise in SWNT-FETs that describes the noise behavior in these devices.

Let us now consider the hysteresis and low frequency noise in bottom-gated SWNT-FETs[19-21]. The bottom-gated SWNT-FETs are not isolated from the environment, resulting in a large hysteresis due to trap charges surrounding the nanotube and its interface with the oxide. The surrounding trap charges affect the effective back gate potential, causing a drift of the threshold voltage. As shown in Fig. 4.6, for SWNT-FETs exposed to the ambient environment, hysteresis is measured in the transfer IV characteristics as a function of the sweeping rate of V_{gs} (Fig. 4.6(a)), the biasing range of V_{gs} (Fig. 4.6(b)), and also the biasing range of V_{ds} (Fig. 4.6(c)). Figure 4.6(a) shows hysteresis in I_d-V_{gs} curves with respect to various sweeping rates for the gate voltage, V_{gs}. The short, medium, and long term V_{gs} scan speed are defined in HP 4516 semiconductor analyzer. A significant dependence of the hysteresis on the V_{gs} scan speed was observed, with slower V_{gs} scans producing larger hysteresis. The result suggests that hysteresis is caused by slow trap charges that charge and discharge on a time scale longer than several seconds. Figure 4.6(b) shows that threshold voltage shift in V_{gs} exhibits a significant dependence on the sweep range of the gate voltage. Gate bias can induce trap charges that can change the threshold voltage. Hysteresis, on the other hand, is independent of V_{ds} as shown in Fig. 4.6(c); confirming that only the threshold voltage is affected by the presence of trap charges. Due to these large hysteresis observed in bottom-gated SWNT, the current fluctuation from traps can dominate the current modulation mechanism. This phenomenon has been observed (but seldom explained) in many reported bottom-gated SWNT-FETs.

The implementation of top-gated SWNT-FETs with ALD Al_2O_3 and HfO_2 allows devices to be isolated from the environment and to show stable DC characteristics. ALD gate oxide layers also serve as a passivation layer for the SWNT and help stabilize the electrical

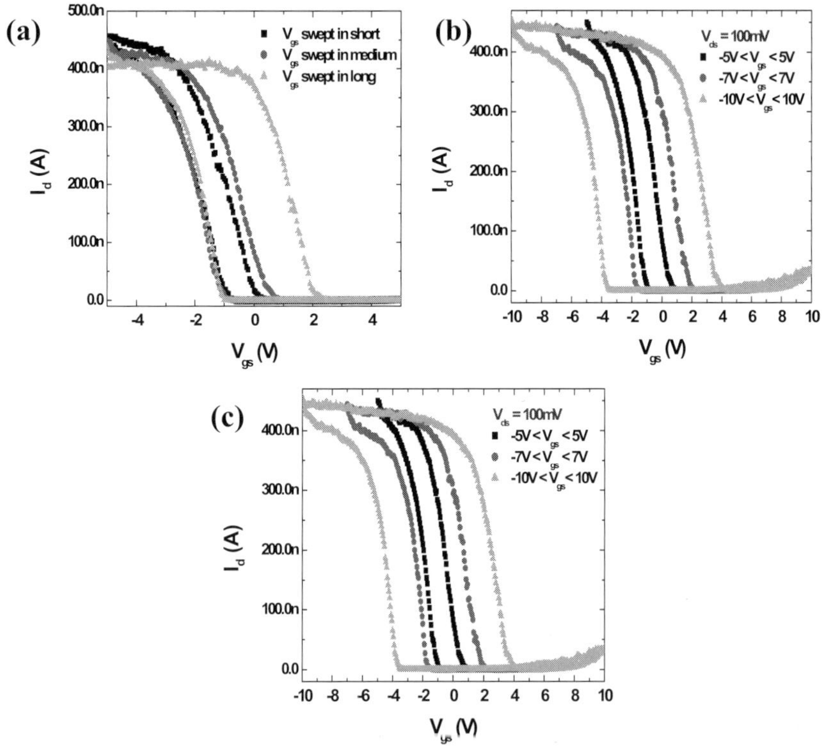

Figure 4.6: Hysteresis behaviors of I_d-V_{gs} curves for SWCNT-FETs under different biasing conditions (a) sweeping speed of gate bias: (squares) fast sweeping rate, (circles) medium sweeping rate, (triangles) slow sweeping rate. (b) I_d-V_{gs} curves for SWCNT-FETs under different gate bias values. Drain bias is fixed at 100 mV. (c) I_d-V_{gs} curves for SWCNT-FETs under different drain voltage as a function of gate bias.

characteristics of SWNT-FETs. Figures 4.3 (b) and (d) show typical I_d versus V_{gs} curves for top-gated transistors. Nearly hysteresis-free characteristics in I_d-V_{gs} curve indicate that ALD Al_2O_3 and HfO_2 suppress trap charges at nanotube-oxide interface.

Similar to hysteresis investigations, low-frequency noise studies can also reveal the existence of trapped charges surrounding the nanotube. Low frequency noise studies can go one step beyond as they can quantify the nature of these traps and how they are affected by the transistor biasing condition. Figure 4.7 shows the correlation between the current noise amplitude (S_I) measured at 100 Hz and drain current (I_d) as a

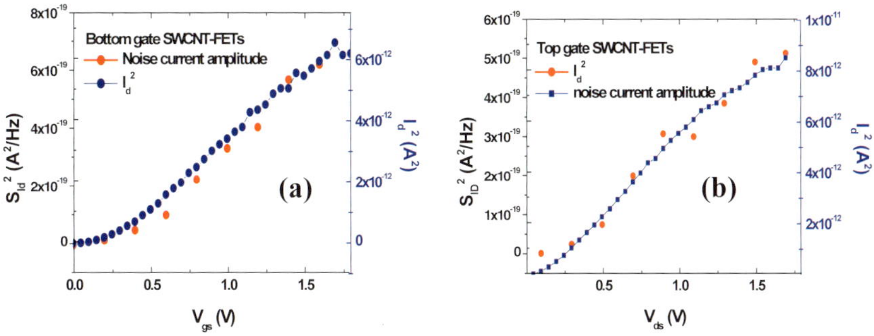

Figure 4.7: Measured low frequency current noise amplitude at 100 Hz for p-type bottom- (a) and top- (b) gated SWCNT-FETs as a function of drain voltage. Gate voltage of -1V and varying drain voltages were applied.

function of the applied voltage V_{gs} for bottom-gated (Fig. 4.7(a)) and top-gated (Fig. 4.7(b)) SWCNT-FETs. Gate voltages of -1 V and 0.6 V were applied to the bottom and top gate contacts, respectively. The 1/f noise spectrum can be found according to Hooge's empirical law[23]

$$S_I(f) = \frac{q\mu_{eff}\alpha_H I\ V}{fL^2} = \frac{q\mu_{eff}\alpha_H I^2}{R_{ch}L^2 f} \tag{2}$$

where α_H is the Hooge parameter, μ_{eff} is the effective mobility of carriers inside the channel and L is the channel length. In both cases (bottom and top gated SWCNT-FETs), current noise amplitude (S_I) is proportional to I_d^2, a trend observed in semiconductor resistors. Figure 4.7 shows that the Hooge parameter of α_H in top gate SWCNT-FETs passivated by ALD HfO_2 is approximately one order of magnitude lower than the one for bottom gated structure. The small noise spectrum of top gated structure indicates that ALD HfO_2 gate oxide can effectively passivate SWNTs and reduce the trap charges at the interface.

Low frequency noise measurements can be used to examine the source of the noise. Assuming a long channel SWNT-FET operating in the linear region,

$$|I_{ds}| = \frac{\mu_{eff}\beta C_g}{L}[(V_{gs} - V_{th})V_{ds} - \frac{1}{2}V_{ds}^2], \quad |g_m| = \frac{\partial I_d}{\partial V_{gs}} = \frac{\alpha_H \beta\mu_{eff} V_{ds}}{fL} \tag{3}$$

where C_g stands for the gate capacitance per number of SWNTs. The total number of carriers, N in the channel is equal to $LI_d/\mu\beta$. By combining Eq. 2 and Eq. 3, the current noise amplitude in the linear regime can be written as

$$S_I(f) = \frac{\alpha_H \beta \mu I_d}{fL} = \frac{\alpha_H \beta \mu_{eff}}{fL}(V_{gs} - V_{th}) g_m \tag{4}$$

The noise amplitude in the linear region is proportional to $g_m(V_{gs} - V_{th})$. In the saturation regime, where $V_{ds} >> V_{dsat} \approx V_g - V_{th}$

$$I_{dsat} = \frac{\mu_{eff} \beta C_g}{2L}(V_{gs} - V_{th})^2, \; g_m = \frac{\partial I_d}{\partial V_{gs}} = \frac{\alpha_H \beta \mu_{eff}}{Lf}(V_{gs} - V_{th}) \tag{5}$$

The total number of carriers (N) is given by

$$N = \int_0^{Vd} -C_g(V_{gs} - V_{th} - V)dV = \frac{C_g}{2}(V_{gs} - V_{th})^2 \tag{6}$$

Combining Eq. 2, Eq. 5 and Eq. 6, the amplitude of current noise in the saturation regime is

$$S_I = \frac{\alpha_H \beta^2 \mu_{eff}{}^2 C_g}{2L^2 f}(V_{gs} - V_{th})^2 = \frac{\alpha_H}{2C_g f}(g_m)^2 \tag{7}$$

This value is proportional to the squared transconductance in the saturation regime. From Eq. 4 and Eq. 7, the amplitude of current noise (S_I) increases with g_m^2 in the saturation regime($|V_{ds}| > |V_g - V_{th}|$) while S_I increases with $(V_{gs}-V_t)g_m$ in the linear regime($|V_{ds}| < |V_g - V_{th}|$). Figure 4.8 shows gate transconductance characteristics (g_m) versus the amplitude of current noise (S_I) at 100 Hz plotted as a function of the gate voltage. For low drain voltages (V_{ds} = -0.2V), the amplitude of the current noise, given by Eq. 4, is proportional to $(V_{gs}-V_t)g_m$. We have observed that the gate voltage at which the channel mobility starts to decrease corresponds to the onset of higher 1/f noise, which increases with $|V_{gs}-V_t|$. Figure 4.8(a) demonstrates that the measured noise behavior in the linear regime corresponds to Eq. 4. Figure 4.8(b) indicates that the intrinsic current noise in the passivated SWNT-FETs with ALD HfO_2 can be represented as a function of transconductance (g_m).

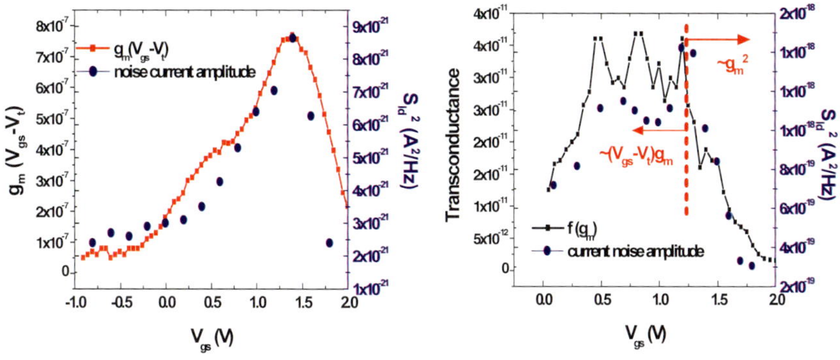

(a) (b)

Figure 4.8: (a) The transconductance characteristics (g_m) versus the amplitude of current noise (S_I) as functions of V_{gs} at the drain voltage of -0.2 V (linear region). (b) The transconductance characteristics (g_m) versus the amplitude of current noise (S_I) as functions of V_{gs} at the drain voltage of -1.5 V (saturation regime).

As shown here, low frequency noise studies can not only indicate the presence of charge trapping centers in the surrounding of nanotubes, they can also identify the behavior of such traps as the device is biased in different operating regimes.

References

1. S. Ijima, *Nature* **354** 56 (1991).
2. S. J. Tans, R. M. Verschueren, and C. Dekker, Nature **393**, 49-52 (1998).
3. T. Durkop, S. A. Getty, E. Cobas, and M. S. Fuhrer, Nano Lett. **4**, 35-39 (2004).
4. A. Javey, K. Hyoungsub, M. Brink, W. Qian, A. Ural, G. Jing, P. McIntyre, P. McEuen, M. Lundstrom, and D. Hongjie, Nature Material **1**, 241-6 (2002).
5. H.-S. P. Wong, J. Appenzeller, V. Derycke, R. Martel, S. Wind, and Ph. Avouris, ISSCC (2003).
6. M. Lundstrom, J. Guo, Springer (2006).
7. Y-M. Lin, J. Appenzeller, and Ph. Avouris, Nano Letters **4**, 947-950 (2004).

8. A. Javey, Q. Wang, A. Ural, Y. Li, and D. Hongjie, Nano Letters **2**, 929-932 (2002).
9. A. Javey, R. Tu, D. B. Farmer, J. Guo, R. G. Gordon, and D. Hongjie, Nano Letters **5**, 345-348 (2005).
10. G. P. Siddons, D. Merchin, J. H. Back, J. K. Jeong, and M. Shim, Nano Letters **4**, 927-931 (2004).
11. A. Bachtold, P. Hadley, T. Nakanishi, and C. Dekker, Science **294**, 1317-1320 (2001).
12. S. J. Wind, J. Appenzeller, R. Martel, V. Derycke, and A. Ph, Applied Physics Letters **80**, 3817-3819 (2002).
13. A. Javey, K. Hyoungsub, M. Brink, W. Qian, A. Ural, G. Jing, P. McIntyre, P. McEuen, M. Lundstrom, and D. Hongjie, Nature Material **1**, 241-6 (2002).
14. A. Javey, J. Guo, D. B. Farmer, Q. Wang, D. Wang, R. G. Gordon, M. Lundstrom, and H.Dai, Nano Lett. **4**, 447-450 (2004).
15. Nathan R.F, Q. Wang, Thomas W, A. Javey, M. Shim, and H. Dai, Appl. Phys. Lett. **81** 913 (2002)
16. C. Kocabas, S. H. Hur, A. Gaur, M. A. Meitl, M. Shim and J. A. Rogers, Small **1** 1110 (2005).
17. H. Stahl, J. Appenzeller, R. Martel, P. Avouris, and B. Lengeler, Phys. Rev. Lett. **85** 5186 (2000).
18. Y.Zhou, A. Gaur; John A.Rogers, Nano Letter **4** 2031 (2004).
19. Y.M. Lin, J. Appenzeller, J. Knoch, Z. Chen, and Ph. Avouris, Nano letters **6**, 930-936 (2006).
20. E. S. Snow, J. P. Novak, M. D. Lay, and F. K. Perkins, Applied Physics letters **85**, 4172-4174 (2004).
21. Philip G. Collins, M. S. Fuhrer, and A. Zettl, Applied Physics Letters **76**, 894-896 (2000).
22. S. Kim et al., "Interface study of single-walled carbon nanotube transistors using low-frequency noise," *65th Device Research Conference* (DRC 07), Notre Dame, IN, June 2007.
23. F. N. Hooge, *Phys.Lett.*, **A-29**, 139 (1969).

5

COOLING WITH INTEGRATED CARBON NANOTUBE FILMS

Géza Tóth, Krisztián Kordás, Pulickel M. Ajayan and Robert Vajtai

The advantageous mechanical and thermal properties of carbon nanotubes (CNTs) in removing heat from hot components is discussed. Experimental results on both CNT-based microfin and brush architectures are described and compared with copper microfin heat sinks. It is shown that thermal management hardware made with CNT structures can be much lighter and more robust than those made from metals. The importance of optimizing the thermal resistance of the solid-solid interface for improving heat transfer is also examined.

5.1 Introduction

The continuously increasing packing density of transistors and the corresponding problem of heat dissipation is a major problem in high performance microprocessor technology. In classical cooling setups, heat-sinks/coolers are integrated on the hot surface or tiny flow channels are formed in the bulk of the component. In both cases the component / cooler geometry, thermal conductivity and the film coefficients at the interfaces of solid and coolant gas/liquid media play crucial roles and need to be optimized for every particular case[1-5]. In future devices, the power densities that should be dissipated are in the range of 100 $W \cdot cm^{-2}$ and even beyond - raising a need to develop novel cooling methods and/or utilize new materials by which heat can be dissipated in a more efficient manner.

Recently, attention has been paid towards new cooling concepts – other then the conventional convective/conductive heat removal techniques. Thermoelectric cooling[6-8] achieved by integrated on-component materials of large Seebeck coefficient and low thermal conductivity exploits the Peltier effect i.e. heat transfer through an interface against a temperature gradient with the consumption of electrical energy. Though the overall efficiency of such a solid-state refrigerator is usually low (~10%), this technique might offer an attractive approach for localized heat removal from microscopic areas.

Ionic wind engines[9,10] represent an innovative and potential method for enhancing heat dissipation in convective coolers, where the local film coefficients of heat transfer are improved by more than 200%. In this method, ionic moieties are generated by field-emitted electrons that collide and change momentum with neutral atoms/molecules thus inducing a gas flow in the vicinity of the surface to be cooled.

Another strategy for improving cooling performance of traditional conductive/convective cooler devices could be implemented by replacing the large finned copper and aluminum blocks with other materials of better thermal conductivity ($\kappa_{Al} \sim 370$ $W \cdot m^{-1} \cdot K^{-1}$, $\kappa_{Cu} \sim 400$ $W \cdot m^{-1} \cdot K^{-1}$). Until lately, diamond was the only known material that could outperform these metals with its outstanding $\kappa_{diamond} \sim 2000$ $W \cdot m^{-1} \cdot K^{-1}$ value, however due to the associated cost it has been applied mainly as a heat-spreader film in electronic components. Engineered anisotropic carbon fiber composites - with a nominal 650 $W \cdot m^{-1} \cdot K^{-1}$ axial thermal conductivity – are promising candidates to replace aluminum and copper heat sinks.[11] The predicted superior thermal conductivity of individual carbon nanotubes ($\kappa_{CNT} \sim 6000$ $W \cdot m^{-1} \cdot K^{-1}$) [12] has also prompted suggestions of their applications in thermal management of high power electronics. With the quick development of synthesis methods,[13-15] various types and forms of CNTs have become available. Therefore, it has been only natural to test their feasibility for thermal management applications. In this chapter, we present a case study to demonstrate practically feasible cooling applications with carbon nanotubes[16].

The excellent thermal conductivity of carbon nanotubes over other existing materials has been exploited in improving the thermal conductivity of nanotube-polymer composites[17-20]. Recently, a micro-channel liquid cooler structure with arrays of carbon nanotubes was used to enhance the heat dissipation of a high power transistor. Using water-flow cooling ~15 $W \cdot cm^{-2}$ enhancement was obtained[21,22].

Cooling of silicon chips utilizing nanotubes integrated directly on the components was reported in reference 16. This otherwise simple method had not been used earlier probably due to the difficulties associated with integrating nanotube structures directly onto chips. This is because the growth temperatures normally encountered during nanotube growth are higher than what can be endured by functional microprocessor silicon elements.

Here, we demonstrate two approaches that exploit the advantageous mechanical[23] and thermal[24] properties of carbon nanotubes in removing heat from hot components. First, we show the cooling of a 1 mm^2 foot-print area chip, where the applied power could be more than 1 W larger when a CNT cooler is applied - compared to the case of the bare chip to reach the same temperature. The CNT fins show similar cooling performance compared to fin structures of similar dimensions fabricated from copper. The assembly uses micro-finned structures of aligned carbon nanotubes, which are mounted onto the back side of a thermometer flip-chip. This chip is fixed on a customized silicon substrate, where we generate Joule heat by an integrated heating circuit. We read out temperature data from the flip-chip (which actually measures its own temperature) and compare this to those recorded for similar assemblies using integrated finned copper coolers as well as just a bare chip without any cooling element. Second, we grow CNT brushes on SiO_2 coated Pt heating wires to help cooling in an efficient manner. Due to the directly anchored CNTs and the as-formed good thermal interface ~100% improvement in heat dissipation could be achieved even through normal convection.

5.2 Experimental

In the experimental work, thick films, consisting of ~1.2 mm long aligned multiwalled carbon nanotubes (MWCNTs) were used (Figure 5.1). The films were grown on Si/SiO_2 templates by catalytic chemical vapor deposition (CCVD) from the precursor of 10 g ferrocene in 1000 cm^3 xylene. The precursor was fed with a rate of 0.1 $cm^3 \cdot min^{-1}$ into an evaporator column pre-heated to 185°C, from which the vapor of

Figure 5.1 a, Low magnification field-emission scanning electron microscopy (FESEM) image showing the micro-structure of three cooler blocks laser-etched next to each other in the MWCNT film. The non-processed slightly tangled layer of ~200 μm thickness at the bottom keeps the fin structure together. Scale bar, 500 μm. b, Grooves and pyramidal fins of aligned nanotubes. Scale bar, 50 μm. c, Close-up image of aligned MWCNTs. Scale bar, 5 μm. The inset in the figure shows energy filtered electron microscopy (EFTEM) image of a well-graphitized nanotube of ~5 nm inner and of ~40 nm outer diameter. Scale bar, 50 nm. Reused with permission from K. Kordás, G. Tóth, P. Moilanen, M. Kumpumäki, J. Vähäkangas, A. Uusimäki, R. Vajtai, and P. M. Ajayan, Applied Physics Letters, 90, 123105 (2007). Copyright 2007, American Institute of Physics.

ferrocene and xylene was introduced into the reactor by argon carrier flow (30 $cm^3 \cdot min^{-1}$, 1 atmosphere). Deposition times of 90-120 minutes were used. The diameters of nanotubes in the films showed a broad distribution of 10-90 nm (Figure 5.1c). The measured bulk density of the nanotube layers was ~0.35 $g \cdot cm^{-3}$·with an approximate nanotube count per unit area of ~100 μm^{-2}.

After detaching the nanotube layers from the templates (in 7:3 vol. % mixture of concentrated HF and ethanol), structures of 10×10 fin array blocks were fabricated in the freestanding films by laser-assisted surface patterning (Figure 5.1 a,b). A defocused (~800 μm offset, spot diameter in the focal plane of ~15 μm) 3ω Nd:YVO_4 pulsed laser (pulse duration of 20 ns, repetition rate of 90 kHz, average power of 300 mW and scan rate of 50 $mm \cdot s^{-1}$ with 15 scan repetitions) was employed to induce localized material removal, thus micro-structuring the nanotube films. The typical size of each finned nanotube array is ~1.2×1.0×1.0 mm^3 with a corresponding mass of ~0.27 mg.

The test structure consisted of an analogue thermometer flip-chip (LM20SITL micro-SMD, by National Semiconductor) mounted on a customized silicon substrate, and an array of MWCNT fin-block soldered on to the backside of the flip-chip (Fig. 5.2). First, the flip-chips were

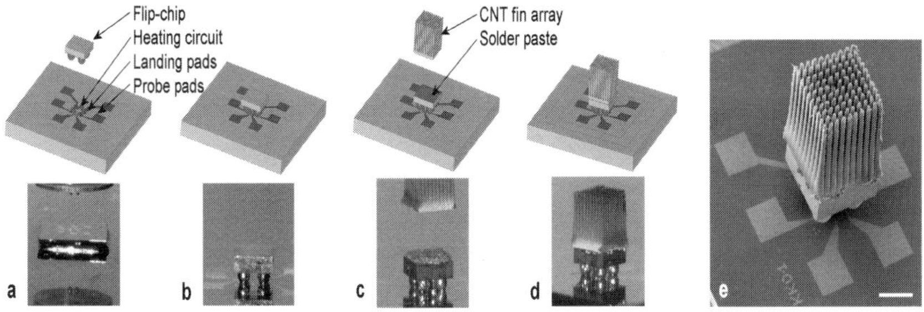

Figure 5.2 Phases of substrate/flip-chip/MWCNT-cooler assembly process: a, Positioning and b, soldering the flip-chip on the Cu landing pads of the substrate. c, Solder paste dispensing, CNT array positioning and d, soldering on the Cu coated back-side of the chip. e, FESEM image of an assembled structure. Scale bar, 500 μm. Reused with permission from K. Kordás, G. Tóth, P. Moilanen, M. Kumpumäki, J. Vähäkangas, A. Uusimäki, R. Vajtai, and P. M. Ajayan, Applied Physics Letters, 90, 123105 (2007). Copyright 2007, American Institute of Physics.

mounted on the substrate, and then the CNT blocks were positioned and soldered on the flip-chips. In order to solder the bottom of the CNT cooler blocks onto the backsides of the chips, both surfaces were cleaned and sputtered with a thin (~50 nm) layer of chromium (adhesion promoter) and subsequently with a solderable copper film of ~2 μm average thickness. (Before sputter coating the backside of the flip-chips were mechanically polished to remove the original coating of an epoxy resin and calcium carbonate composite.) The solder bumps (63Sn-37Pb, eutectic, $T_m = 183°C$) of the polished and metallized flip-chips were dipped in a solder flux, and then the chips were aligned and soldered at 230°C to the Cu landing pads of the substrates using a split-field microscope equipped with a hot-plate. In the next step, each sample was tested on a probe-station to measure the chip temperature vs. heating power characteristics under various thermal loads (up to ~7 W) and cooling gas flow rates (N_2 @ 0-2 L·min^{-1} through a cylindrical nozzle of 0.44 mm inner diameter). The temperature of the heated chips was extracted from the output analogue (voltage) data of the thermometer chips we used in the assemblies.

After testing, a thin (~10 μm) layer of low melting point eutectic 46Bi-34Sn-20Pb solder paste ($T_m = 96°C$) was dispensed on the back of the chip, and using a fine-placer a block of nanotube cooler along its Cr/Cu sputtered side was positioned and soldered onto the chip at 170°C. In the course of the soldering process, the finned array was pushed against the chip to squeeze out the excess solder melt. Finally, the assemblies were tested again using the same routine as in the first phase.

5.3 Results and Discussion

In the experiments described here, Joule heat was generated on the substrate by an integrated heating circuit. The nanotube-on-chip assembly showed a reduced chip temperature as compared to the corresponding chip without the cooling block. With natural convection, ~11% more power was dissipated from the chip that has the attached nanotube fin structures - with a corresponding figure of merit of 31.9 $mW \cdot K^{-1}$ instead of 28.8 $mW \cdot K^{-1}$ (Figure 5.3). When applying forced N_2 flow, the cooling efficiency improved by ~19%; i.e. 82.8 $mW \cdot K^{-1}$ of power could be dissipated instead of 69.8 $mW \cdot K^{-1}$. Accordingly, the finned CNT structures would allow the dissipation of ~30 $W \cdot cm^{-2}$ and ~100 $W \cdot cm^{-2}$ more power at 100°C from a hot chip for the cases of natural and forced convection, respectively. Because of the low density of CNT films[24], the weight normalized additional power dissipation of such structures was estimated to be as high as 1.1 $kW \cdot g^{-1}$ or 3.7 $kW \cdot g^{-1}$ for natural and forced convection (as calculated for the corresponding geometry from the measured density of the nanotube films, ~0.35 $g \cdot cm^{-3}$).

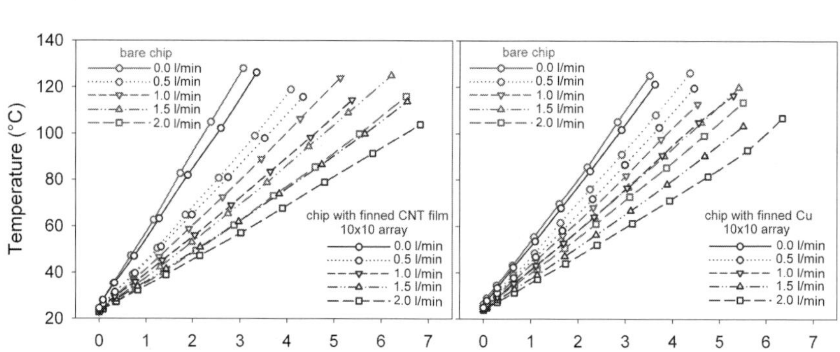

Figure 5.3 Measured temperatures for reference, CNT- and Cu-equipped chips with 10×10 fin arrays. The graph in the left panel shows chip temperatures measured for chip-on-substrate (red plots) and for the corresponding MWCNT 10×10 fin array-on-chip-on-substrate assemblies (blue plots) under various heating powers and N_2 flow rates. The graph in the right panel shows the results of a comparative experiment in which a finned copper cooler (10×10 array) was used.

These figures project the possibility of such light-weight solid state add-on structures for on-chip thermal management without involving complex fluid flow for heat removal. To compare the cooling performance of the micro-structured CNT films with other materials commonly used in thermal management, finned copper blocks with similar geometry to that of the CNT structures were fabricated and mounted on test chips. The measured temperature *vs.* power curves show very similar results for the two materials, i.e. copper and nanotubes perform fairly equally.

Experiments with mounted CNT blocks without a fin structure revealed poor cooling performance. The dense forest of nanotubes ($\sim 10^2\ \mu m^{-2}$) impedes the coolant flow in the films, limiting the heat exchange only to the upper facet of the films. In contrast, the grooves in the structured CNT films enable better coolant flow and heat dissipation from the fins towards the surrounding medium via convective fluxes.

Computational fluid dynamics (CFD) and thermo-electric finite element models were applied to study the steady-state laminar coolant flow (with Reynolds number of ~1800 at the nozzle exit for 0.5 l/min flow rate) and also the temperature distribution for both the finned and bare reference chips. The computed flow velocity fields were then used to approximate the corresponding average heat transfer coefficients h, at different locations of the experimental geometry, which were subsequently used in the heat-flow calculations. Depending on the local flow rates, the estimated values for h varied from 20 $W \cdot m^{-2} \cdot K^{-1}$ up to 500 $W \cdot m^{-2} \cdot K^{-1}$. These agree well with those published for impingement gas flows[25,26].

Because of the relatively low temperatures and small surface areas of the hot spots, the losses caused by the heat radiation were neglected in the thermal model, i.e. the heat transport was restricted to conductive and convective fluxes. The heat-flow model, in which we used flow dependent film coefficients and temperature dependent materials parameters, gave good agreement with the chip temperatures we measured for various experimental conditions (Figure 5.4).

The simulations revealed a by-pass flow around the cooler block besides the desired side-way flow trough the channels between the fins. To study the role of the otherwise unwanted by-pass flow, a new structure consisting of 6×6 fin-array (instead of the original 10×10) was made (Figure 5.5). Despite the enhanced coolant flow through the 6×6 array, the measured cooling efficiency with such coolers (both CNT and

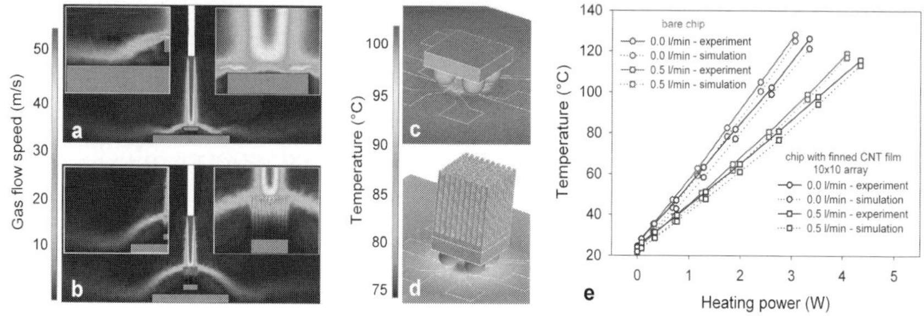

Figure 5.4 Steady-state solutions of computer simulations (finite element modeling) for coolant flow velocity and temperature distribution for a chip without and with a finned cooler. Panels a, and b, show color plots for the coolant velocity for a chip without and with a finned cooler, respectively. Insets show more detailed plots of the areas at the close proximity of the surfaces. Panels c, and d, show color plots of the temperatures for a chip without and with a finned cooler, respectively. Heating power of 3 W for each case was applied for 0.5 l/min coolant flow. The fins were considered as a bulk medium with an effective density of 350 kg·m^{-3}, and with an effective anisotropic thermal conductivity of 300/30/30 W·m^{-1}·K^{-1}. Specific heat of 650 J·kg^{-1}·K^{-1} was used. The graph in panel e, compares the experimental and computational results for the power vs. temperature curves for both natural convection and forced convection (0.5 l/min) conditions. Reused with permission from K. Kordás, G. Tóth, P. Moilanen, M. Kumpumäki, J. Vähäkangas, A. Uusimäki, R. Vajtai, and P. M. Ajayan, Applied Physics Letters, 90, 123105 (2007). Copyright 2007, American Institute of Physics.

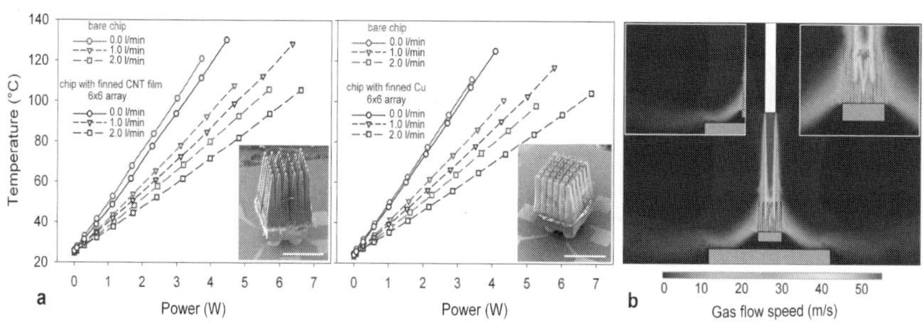

Figure 5.5 Measured temperatures for reference, CNT- and Cu-equipped chips with 6×6 fin arrays and the corresponding flow field of coolant gas. a, The graph in the left panel shows chip temperatures measured for chip-on-substrate (red plots) and for the corresponding MWCNT 6×6 fin array-on-chip-on-substrate assemblies (blue plots) under various heating powers and N_2 flow rates. The graph in the right panel shows the results of a comparative experiment in which a finned copper cooler (6×6 array) was used. b, Simulated flow field of coolant showing good gas impingement in the groves between the fins.

Cu were tested) showed only minor improvement compared to the original 10×10 fin-arrays. This result suggests the presence, and significant effect, of a thermally resistive interface between the chip and the coolers. For realizing coolers with higher efficiency one needs to optimize not only the fin geometry, but even more importantly the thermal resistance of the solid-solid interface needs to be minimized with improved engineering and assembly of the components[27,28].

Beyond the earlier mentioned advantageous mechanical and thermal properties as well as light weight, CNTs also dominate other materials from the point of view of ease of direct fabrication and integration onto complex surfaces. To demonstrate direct on-component carbon nanotube architectures – which might be essential for future cooling devices – we grew bundles of MWCNT brushes (length of ~500 μm) on Pt wires (diameter of 25.4 μm) coated with a thin layer of SiO_2. The temperature *vs.* heating power characteristics of pristine Pt/SiO_2 and also MWCNT-

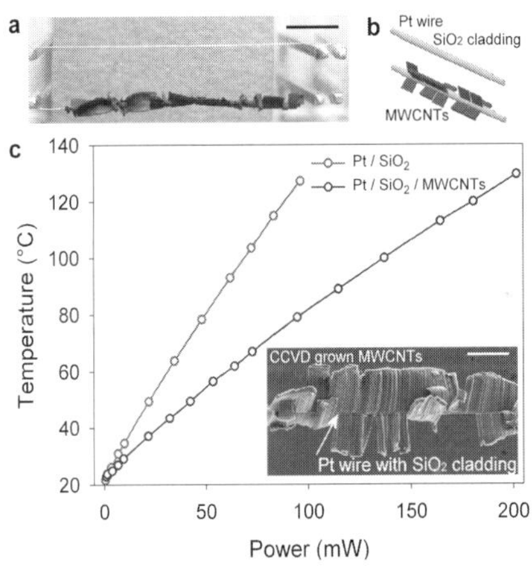

Figure 5.6 Thermal management with carbon nanotubes grown directly on a heater (Pt wire coated with SiO_2). a, Optical image of two platinum wires of 25.4 μm diameter and ~16 mm length. Both wires were coated with PECVD grown SiO_2 (Pt/SiO_2), but the lower one is equipped with MWCNTs of ~500 μm length (Pt/SiO_2/MWCNTs) grown directly on the SiO_2 coating. Both wires were suspended and soldered in their cleaned metal end to metal electrodes to enable electrical current through the Pt core. Scale bar, 3 mm. b, Schematic drawing of the two types of wires. c, Temperature vs. power plots for the Pt/SiO_2 and Pt/SiO_2/MWCNTs wires having R_0 resistances (Pt core) of 3.73 Ω and 3.48 Ω, respectively. The temperature of the Pt cores was calculated from the temperature dependent change of relative resistance. The inset shows a FESEM image taken from the CNT bundles grown on the SiO_2 coated Pt wire. Scale bar, 500 μm.

decorated wires were measured and compared. The experiments revealed very efficient heat dissipation for samples with nanotube bundles with an improvement of ~100% under conditions of normal convection (Figure 5.6). This significant enhancement of cooling performance is due to the proximity of the cooler to the heat source and also because of the increase in the exposed surface area.

5.4 Conclusions

Cooling of silicon chips and hot wires with integrated MWCNT films are demonstrated. Application of nanotubes offers a light-weight, effective, and facile alternative of larger and much heavier metal block coolers or liquid-coolant based complex systems. Owing to the individual and collective mechanical strengths in spite of the severe mechanical load conditions during the assembly and testing phases of the experiments, no any damage or degradation of the nanotube films was observed proving the robustness and feasibility of CNT cooler structures for practical applications. The demonstrated assembly techniques use conventional manufacturing methods thus providing an easy protocol to transfer and integrate nanotube arrays onto presently used platforms. Tailoring nanotube structure to obtain higher thermal conductivity, improvement and optimization of the chip-nanotube thermal interface, fin-array geometry/location, nanotube forest density, and improved gas flow efficiency etc. could lead to additional enhancement of the ~100 $W \cdot cm^{-2}$ power dissipation reported here.

Acknowledgements

We acknowledge financial support from the Interconnect Focus Center New York at RPI, an FCR program of the Semiconductor Research Corporation. R.V. acknowledges the NSEC Initiative of the NSF under Award DMR-0117792. G.T. thanks for the grants received from Nokia Foundation and from the Tekniikan edistämissäätiö. K.K. is grateful for the Academy of Finland for the research fellow post and also for the incentive funding received (120853). We acknowledge the assistance of Mia Kumpumäki and Pekka Moilanen; and also the technical support from the Micro- and Nanotechnology Center, University of Oulu.

References

1. A. Bar-Cohen and W. M. Rohsenow, *Thermally optimum spacing of vertical, natural convection cooled, parallel plates.* J. Heat Transf. ASME **106**, 116 (1984)

2. R. E. Simons, V. W. Antonetti, W. Nakayama and S. Oktay, *Heat transfer in electronic packages* in Microelectronics Packaging Handbook, Ch. 4 (Chapman & Hall, New York, 1997)

3. D. B. Tuckerman and R. F. W. Pease: *High-performance heat sinking for VLSI*, IEEE Electron Device Lett. **2**, 126 (1981)

4. M.J. de Boer, R. W. Tjerkstra, J.W. Berenschot, H.V. Jansen, G.J. Burger, J.G.H. Gardeniers, M. Elwenspoek and A. van den Berg: *Micromachining of buried micro channels in silicon*, J. Microelectromech. Syst. **9**, 94 (2000)

5. Y. Peles, A. Kosar, C. Mishra, C.-J. Kuo and B. Schneider: *Forced convective heat transfer across a pin fin micro heat sink*, Int. J. Heat Mass Transfer **48**, 3615 (2005)

6. T. C. Harman, P. J. Taylor, M. P. Walsh and B. E. LaForge, *Quantum dot superlattice thermoelectric materials and devices,* Science **297**, 2229 (2002)

7. Y. Zhang, A. Shakouri and G. Zeng, *High-power-density spot cooling using bulk thermoelectrics*, Appl. Phys. Lett. **85**, 2977 (2004)

8. P. Wang and A. Bar-Cohen, *On-chip hot spot cooling using silicon thermoelectric microcoolers,* J. Appl. Phys. **102**, 34503 (2007).

9. M. Molki and K.L. Bhamidipati, *Enhancement of convective heat transfer in the developing region of circular tubes using corona wind,* Int. J. Heat Transfer **47**, 4301 (2004)

10. D. B. Go, S. V. Garimella, T. S. Fisher and R. K. Mongia, Ionic winds for locally enhanced cooling, J. Appl. Phys. **102**, 053302 (2007).

11. U.S Patent, No. 6.844.054

12. M. S. Dresselhaus, and P. C. Eklund, *Phonons in carbon nanotubes,* Adv. Phys. **49**, 705-814 (2000)

13. A. Y. Cao, V. Veedu, X. S. Li, Z. Yao, M. N. Ghasemi-Nejhad and P. M. Ajayan, *Multifunctional brushes made of carbon nanotubes,* Nature Mater. **4**, 540 (2005)

14. X. S. Li, A. Y. Cao, Y. J. Jung, R. Vajtai and P. M. Ajayan, *Bottom-up growth of carbon nanotube multilayers: Unprecedented growth.*, NanoLett. **5**, 1997 (2005)

15. S. Talapatra, S. Kar, S. K. Pal, R. Vajtai, L. Ci, P. Victor, M. M. Shaijumon, S. Kaur, O. Nalamasu and P. M. Ajayan, Direct growth of aligned carbon nanotubes on bulk metals, Nature Nanotech., **1**, 112 (2006).
16. K. Kordas, G. Toth, P. Moilanen, M. Kumpumäki, A. Uusimäki, J. Vähäkangas, R. Vajtai, P.M. Ajayan, *Chip cooling with integrated carbon nanotube microfin architectures*, Appl. Phys. Lett. **90**, 123105 (2007)
17. M. J. Biercuk, M. C. Llaguno, M. Radosavljevic, J. K. Hyun, A. T. Johnson, and J. E. Fischer, *Carbon nanotube composites for thermal management*, Appl. Phys Lett. **80**, 2767 (2002)
18. Y. Wu, C. H. Liu, H. Huang and S. S. Fan, *Effects of surface metal layer on the thermal contact resistance of carbon nanotube arrays*, Appl. Phys. Lett. **87**, 213108 (2005)
19. C-W. Nan, Z. Shi and Y. Lin, *A simple model for thermal conductivity of carbon nanotube-based composites,* Chem. Phys. Lett. **375**, 666 (2003)
20. J. Hone, M. C. Llaguno, M. J. Biercuk, A. T. Johnson, B. Batlogg, Z. Benes and J. E. Fischer, *Thermal properties of carbon nanotubes and nanotube-based materials,* Appl. Phys. A **74**, 339 (2002)
21. M. Zhimin, R. Morjan, J. Anderson, E. E. B. Campbell and J. Liu, *Integrated nanotube microcooler for microelectronics applications*, IEEE 55th Electronic Components and Technology Conference, Lake Buena Vista, Fl, 31 May 2005. Proc., pp. 51-54 (2005)
22. M. Zhimin, J. Anderson and J. Liu, *Integrating nano carbontubes with microchannel cooler*, IEEE 6th High Density Microsystem Design and Packaging and Component Failure Analysis Conference, Shanghai, China, 30 June 2004. Proc., pp. 373 – 376 (2004)
23. A. Y. Cao, P. L. Dickrell, W. G. Sawyer, M. N. Ghasemi-Nejhad and P. M. Ajayan, *Super-compressible foamlike carbon nanotube films*, Science **310**, 1307 (2005)
24. P. Kim, L. Shi, A. Majumdar and P. L. McEuen, *Thermal transport measurements of individual multiwalled nanotubes,* Phys. Rev. Lett. **87**, 215502 (2001)
25. L. A. Brignoni and S. V. Garimella, *Experimental optimization of confined air jet impingement on a pin fin heat sink*, IEEE Trans. Comp. Packag. Manufact. Technol. **22**, 399 (1999)

26. H. A. El-Sheikh and S. V. Garimella, *Enhancement of air jet impingement heat transfer using pin-fin heat sinks*, IEEE Trans. Comp. Packag. Manufact. Technol. **23**, 300 (2000)
27. J. Xu and T. S. Fisher, *Enhancement of thermal interface materials with carbon nanotube arrays*. Int. J. Heat Mass Transfer **49**, 1658 (2006)
28. S. T. Huxtable, D. G. Cahill, S. Shenogin, L. R. Ozisik, P. Barone, M. Usrey, M. S. Strano, G. Siddons, M. Shim and P. Keblinski, *Interfacial heat flow in carbon nanotube suspensions,* Nature Mater. **2**, 731 (2003)

6

AC DIELECTROPHORESIS ALIGNMENT OF GALLIUM NITRIDE NANOWIRES (GaN NWs) FOR USE IN DEVICE APPLICATIONS

Sang-Kwon Lee, Tae-Hong Kim and Seung-Yong Lee

Simple and effective ac and dc dielectrophoresis (DEP) methods to align and manipulate semiconductor gallium nitride (GaN) nanowires (NWs) with variations of the types of electrical fields as well as frequency are presented in this chapter. The ability of the alignment and the formation of the nanowire assembly (single or bundle configuration) strongly depends on the magnitude of both ac and dc electric fields. The yield results indicate that the GaN NWs using ac DEP are much more effective with a high yield of ~80% over the entire array in the chip compared to that using dc DEP. In addition, simple hybrid p-n junction structures, which are fabricated by the coupling of n-GaN NWs and p-GaN substrates, have bright UV-blue emission (411-437 nm) from the n-GaN NW/p-GaN substrate junction. The light emission is strong enough to be observed with the naked eye even with a single GaN NW LED. Finally, a DEP-prepared GaN nanowire half-wave rectifier (n-GaN NW/p-Si substrate) is also described.

6.1 Introduction

1-D semiconductor nanowires (NW) offer a good system to investigate the dependence of the electrical, thermal and mechanical properties on dimensionality and size reduction or quantum confinement effects.[1,2] They are very interesting building blocks for the fabrication of various devices on a nanoscale range due to their own superior properties such as the capability for doping modulation and the selectivity of the carriers to those of the carbon nanotube (CNT). In addition, they can be synthesized using diverse techniques such as laser-assisted catalytic growth, chemical vapor deposition (CVD) hydride vapor phase epitaxy (HVPE), and metal-organic chemical vapor deposition (MOCVD)[1-5]. The ability to manipulate and align these individual nanostructures is necessary for wafer-based large scale integration. It also helps with proper electronic device applications and the characterization of their electrical properties.

Dielectrophoresis (DEP) has been utilized to manipulate nanoparticles, carbon nanotubes, metallic nanowires, and semiconductor nanowires such as indium phosphide (InP), tin oxide (SnO_2), and zinc oxide (ZnO) nanowires[6-13]. Duan *et al* (2001) first demonstrated the InP nanowire p-n junction structures exhibiting electroluminescence and photoluminescence using a fluidic assembly method. They reported the possibility of the electric-field assisted assembly technique (dc dielectrophoresis) to align and control the nanowires[14]. However, there are few publications concerning systematic studies of dielectrophoresis related to device applications with semiconductor nanowires.

In this chapter, we present systematic investigations of the DEP process to align semiconductor gallium nitride nanowires (GaN NWs) across pairs of round metal electrodes with variation of the type of electrical field and the frequency. We also present high-brightness GaN nanowire UV-blue light emitting diodes (LEDs) which were fabricated by the coupling of n-GaN nanowires and p-GaN substrates using dielectrophoresis techniques. Furthermore, we'll also show the electrical characteristics of these UV-blue GaN NW LEDs. In addition, we also demonstrate GaN nanowire (NW) half-wave current rectifiers which were formed on a patterned p-Si substrate. Finally, some results of material characterization using techniques such as X-ray diffraction (XRD), photoluminescence (PL), and field-emission scanning electron microscopy (FE-SEM) are also described.

6.2 Sample Preparation

6.2.1 GaN synthesis

Single crystal gallium nitride nanowires (GaN NWs) were prepared by hot-wall chemical vapor deposition (CVD). C-plane sapphire wafers were used as substrates for the GaN NW growth. The substrates, deposited on a 2 nm layer of Ni by sputtering were placed in a uniform temperature region in a quartz tube reactor. Gallium (Ga) and nitrogen (N) components are supplied to the substrate by using metallic Ga and NH_3 gas. Typically, the system was heated to ~900°C under a flow of NH_3 at a rate of 20 $cm^3 \cdot min^{-1}$ and maintained for 6 hrs, and then cooled down to room temperature[15-18].

6.2.2 Dielectrophoresis (DEP)

For the DEP experiments, GaN nanowire suspensions were prepared by sonicating a 10mL IPA (Isopropyl Alcohol) mixture. Then metal electrodes (Ti/Au=50/100 nm) were deposited with a standard photo-lithography process on a 4 inch diameter thermally oxidized (400 nm) silicon wafer (n-type, 100 orientation, resistivity 5 – 10 Ωcm). One chip (5×5 mm^2) consists of 200 opposing pairs of electrodes whose geometry includes a 4 μm gap between two Ti/Au electrodes. Fig 6.1 shows images of the dielectrophoresis measurement setup and that of the metal electrodes used for the DEP experiments (200 pairs).

A drop of the GaN NW suspension (~ 3μL) is placed on the selected gap using a micropipette while the electrical field is being applied across the electrodes (one of them grounded and the other applied with dc or a sinusoidal ac voltage) as shown in Fig 6.1 (b). The electric field was continuously applied until the suspension completely dried out. Then the samples images were taken by either a scanning electron microscope (SEM) or an optical microscope in order to characterize the morphology and the yield of the aligned nanowires between the electrodes. The yield rate, the percentage of the number of metal electrodes where at least one GaN nanowire aligns out of 200 electrodes was calculated by counting the GaN nanowires aligned in the gap across the electrodes. For the DEP experiment, 14 samples (D1 ~ D14) were prepared differently. As shown in Table 6.1, the first 4 samples (D1~D4) and the next 10 samples

(D5~D14) were subjected to applied dc (1, 5, 15, 20 V) and ac electrical fields (1,5,10,15,20 V_{p-p}) across the metal electrodes, respectively. Two frequencies (10 kHz and 20 MHz) were selected to investigate the effect of the modulating frequency on the DEP experiment.

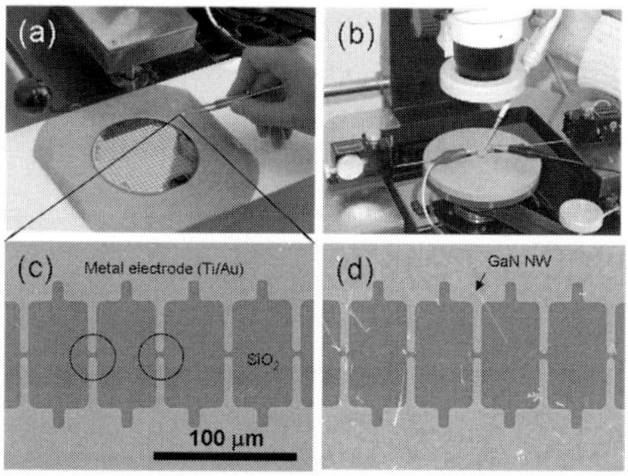

Fig 6.1 (a), (b) Images of the dielectrophoresis (DEP) measurement set-up connected to a function generator and a power supply for supplying the ac and dc electric fields. Scanning electron microscope (SEM) images of opposing pairs of the Ti/Au electrodes (c) before and (d) after dielectrophoresis. The circles denote the electrode gap where the GaN nanowires are aligned.

Table 6.1 Summary of the sample preparation for all samples (D1 D14)

	Samples	Frequency	DC Voltage (V)	AC Peak-Peak Voltage (V_{p-p})
DC	D1~D4	-*	1(D1),5(D2), 15(D3), 20(D4)	-*
AC	D5~D9	10 kHz	-	1(D5),5(D6),10(D7),15(D8),20(D9)
	D10~D14	20 MHz	-	1(D10),5(D11),10(D12),15(D13),20(D14)

* not performed

6.2.3 UV-blue GaN NW LEDs

For LED structures, the p-n junction diodes were fabricated by coupling n-GaN nanowires together with a p-GaN substrate. First, a 200 nm-thick oxide was deposited on a p-GaN substrate (0.8×0.8 cm^2, 1 μm thick), which was prepared on a c-plane sapphire substrate using a metal-organic chemical vapor deposition (MOCVD)[17] process. The resistivity, the mobility, and the carrier concentration of the p-GaN substrate were measured with a Hall effect measurement setup to be 3.28 Ωcm, 11.67

cm^2/Vs, and 1.6 $\times 10^{17}$ cm^{-3}, respectively. Circular-shaped anode contacts (Ni/Au=30/150 nm, 100 μm diameter) were defined on p-GaN substrates after etching the oxide by dipping in dilute buffered hydrofluoric (BHF) acid for 5 min. On the other hand, the donut-shaped cathode metal electrodes (Ni/Au=30/150 nm, 500 μm diameter) for contact to n-type GaN nanowires were patterned on the oxide by a standard photo-lithography process. All of the samples were annealed at 650 – 750°C for 30 s for achieving lower contact resistance to GaN. One sample (0.8×0.8 cm^2) consisted of 25 donut-shaped p-n junction LED structures, whose geometry included a 12 μm gap between two electrodes (anode and cathode). The p-n junction (n-GaN nanowire/p-GaN substrate) can be obtained in two different ways: random dispersion and dielectrophoresis technique. Here, the dielectrophoresis assisted assembly deposition (DAAD)[17] technique will be described.

6.3 Results and Discussion

6.3.1 Material characteristics

Gallium nitride nanowires (GaN NWs) were prepared by conventional hot-wall CVD technique[15-17]. Figs 6.2 (a) and (b) show scanning electron

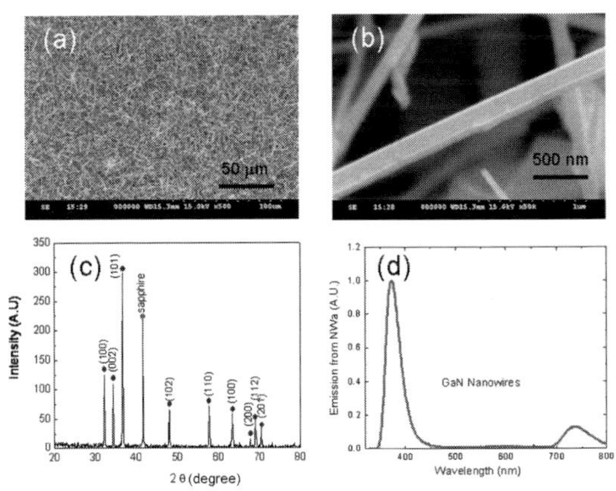

Fig 6.2 (a) Scanning electron microscope (SEM) image of GaN nanowires grown on a c-plane sapphire substrate with enlarged image (b). (c) X-ray diffraction (XRD) spectra of GaN nanowires. (d) Photoluminescence (PL) spectrum of GaN nanowires measured at room temperature. See Refs. [15-16].

microscope (SEM) images of GaN nanowires prepared on the c-plane sapphire substrate. The GaN NWs were distributed over the entire area of the sapphire substrate (8×8 mm^2) and had diameters of > 100 nm with length of several μm. X-ray diffraction (XRD) pattern of the GaN NWs is indexed to a hexagonal-based wurtzite structure as shown in Fig 6.2 (c). Photoluminescence (PL) emission peaks shown in Fig 6.2 (d) were observed at the center wavelengths of 374 nm (3.31 eV in photon energy) with a He-Cd CW laser excitation source at room temperature.

6.3.2 Dielectrophoresis characteristics

Figs 6.3 (a)-(d) show the SEM images of aligned GaN nanowires on the patterned electrodes and the yield of the aligned GaN nanowires in the gaps over 200 opposing electrodes with increasing ac and dc electric fields. In Fig 6.3 (d) the results indicate that the alignment yield of the GaN nanowires strongly depends not only on the dc electric field but also on the ac electrical field although it has been observed that the yield of GaN NWs does not increase after 15 V for dc electric field application. Generally, it was observed that the yield rate increases with both increasing dc and ac voltages. To understand such behavior, observed in GaN nanowires under applied electric field, we should consider the well-known standard model of DEP exerted on nanowires. For a homogeneous cylindrical shape and a long nanowire with its major axis parallel to an inhomogeneous alternating electric field, the DEP force is given by[19-20]

$$\vec{F}_{DEP} = \frac{v}{2}\varepsilon_m K(\omega)\vec{\nabla}\left(\vec{E}_{rms}^2\right) = \frac{\pi^2 r^2 l}{2}\varepsilon_m K(\omega)\vec{\nabla}\left(\vec{E}_{rms}^2\right) \quad (1)$$

Where the v is the volume of the nanowires, $K(\omega)$ is the real part of the Clausius-Mosotti factor, r is the radius of the nanowires, and l is the length of the nanowires. The $K(\omega)$ is related to the nanowire dielectric constant ε_n and liquid medium dielectric constant ε_m by

$$K(\omega) \equiv \mathrm{Re}\left[\frac{\varepsilon_n^* - \varepsilon_m^*}{\varepsilon_m^*}\right] \quad (2)$$

Here the asterisk * denotes that the dielectric constant is a complex quantity, and that it can be related to the conductivity σ and the angular frequency ω through the standard formula, $\varepsilon^{*} = \varepsilon - i(\sigma/\omega)$. Eq. (1) clearly indicates that the DEP force depends strongly on the volume of the nanowire, Clausius-Mosotti factor, and the gradient of the electric field. Dong et al.[8] also reported that the alignment of the nanotubes is not only controlled by the DEP force but also by the torque $\vec{T}$ exerted on the induced electrical dipole moment. Torque is given by

$$\vec{T} = q\vec{d} \times \vec{E} \tag{3}$$

Where q is the induced electrical charge on the nanowires, $\vec{d}$ is the displacement between the induced charges, and $\vec{E}$ is the electrical field.

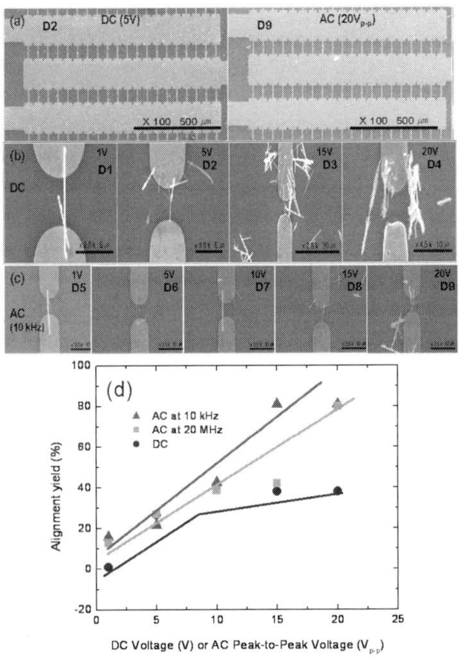

Fig 6.3 SEM images of aligned GaN nanowires on the patterned 200 electrode array. (a) The general shape and the morphology of the chip for the sample D2 (dc 5V) and D9 (ac 20V$_{p-p}$) after the dielectrophoresis process. Aligned GaN nanowires in the gap across the electrode (b) for the samples D1, D2, D3, D4 and (c) for the samples D5, D6, D7, D8, D9. (d) the yield of the aligned GaN nanowires in the gap over 200 opposing electrodes as a function of dc and ac electric field. See Refs. [15-16].

As shown in Fig 6.3 (a), when the dc electric field was applied to the electrodes of the sample D2, many GaN nanowires, including some metallic impurity particles which were not aligned across the electrodes, were deposited on the charged electrodes. These prevented the deposition of more than one nanowire across the electrodes. On the other hand, we did not observe any impurities and nanowires on the electrodes with ac electric field application (sample D9), as shown in Fig 6.3 (a). Most of the nanowires were aligned in the gap across the electrodes and the edge of the electrodes and not on the electrodes themselves as seen for dc electric field. From our measurements ac dielectrophoresis is clearly superior to dc dielectrophoresis in terms of the alignment of the nanowires. For dc and ac electric fields it was observed that the number of the aligned nanowires across the electrodes increased with the applied dc and ac voltages. From Fig 6.3 (b) and (c) we observe that the morphology of the aligned GaN nanowires in the gap across the electrodes changes from the single to the bundle type of nanowires. Optimum DEP process for the alignment of GaN nanowires with a single nanowire over the electrode array is from the ac electric field DEP process with a voltage amplitude below $10V_{p\text{-}p}$ at a frequency in the range of 10 kHz - 20 MHz. For the electric field dependence shown in Fig 6.3 (d), by increasing the dc (up to 20 V) and ac (up to 20 $V_{p\text{-}p}$) electrical field, we could achieve a high yield of aligned GaN nanowires in the electrodes. This can be simply explained by a high dc and ac dielectrophoresis force with a larger electric field as predicted in Eqs. (1) and (3). The larger electric fields (both dc and ac) forced the GaN nanowires to align across the electrodes such that the assembly yield reached around 80% over the entire electrode array in the chip. In order to understand the frequency dependence of the yield rate, we should notice the value of the $K(\omega)$ factor. For GaN nanowires within the IPA liquid medium, the theoretical calculation of $K(\omega)$ at different angular frequencies is bounded by the limits $1.7\times10^{7} < K(\omega/2\pi) < 1.0$ in the frequency range of 1 kHz - 80 MHz. We take it here that the conductivity and dielectric constant for GaN nanowires and IPA are σ_n=104 S/m, ε_n=12.2 ε_0 and σ_m =6×10^{-6} S/m, ε_m =18.3 ε_0, respectively.[2,21-23] The sign of $K(\omega)$ denotes the direction of the electric field. Thus, in our case the nanowires at the two frequencies of 10 kHz and 20 MHz were attracted to the electrode edges since the values of $K(\omega)$ for these two frequencies are positive (known as positive DEP).[19] In Fig 6.3, the alignment yield for 10 kHz is slightly higher than that for 20 MHz. This could be a result of the magnitude of $K(\omega)$ difference at the two frequencies in producing the DEP force.

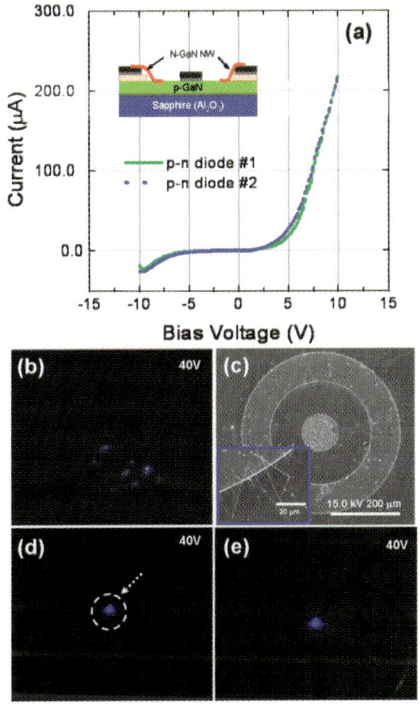

Fig 6.4 (a) Current-voltage (I-V) characteristics of random dispersion prepared GaN nanowire LEDs. The measurements are performed with two GaN nanowires LEDs located in different regions on the sample chip (0.8 x 0.8 cm^2), consisting of 25 LEDs. Electroluminescence (EL) properties for aligned GaN nanowire LED prepare by dielectrophoresis techniques. EL (b) and SEM image (c) from aligned GaN nanowire. Inset in (c) shows the aligned GaN nanowires around the cathode electrodes. EL images taken by digital camera without and with background light from the GaN nanowires are shown in (d, e). See Ref. [17].

6.3.3 GaN NW LEDs

Current-voltage (I-V) measurements shown in Fig 6.4 (a), show well-defined current rectifying behavior, as expected from homojunction p-n diodes with a turn-on voltage of ~ 3.4 V. Little leakage current and no breakdown were observed for reverse bias up to -25 V. The observed reverse leakage current was ~ 5×10^{-4} A at 25 V of reverse bias voltage at room temperature. To illustrate the reproducibility of the n-GaN nanowire/p-GaN substrate junction LED structures, 25 samples were fabricated with the same process and we found that all of the junctions exhibited similar rectifying behavior and functioned with low leakage currents. The on-state series resistances (R_S) for these GaN NW p-n junction LEDs were determined to be in the range from 20 kΩ to 25 kΩ, from the inverse of the slope in the I-V curve in Fig 6.4 (a). To maximize light emission from single GaN nanowire LEDs and to meet the requirements for wafer-based large-scale integration, GaN nanowires were assembled across counter electrodes (anode and cathode) by dielectrophoresis assisted assembly deposition (DAAD)[15-19]. In Fig 6.4 (b)-(e), we observe strong UV-blue light emission, which can be readily observed with the naked eye. The results indicate that the light emissions could be intensified with the formation of multiple GaN nanowires junctions operated at 40V, as is seen in Fig 6.4 (b), (d) and (e). Previous studies on single crossed GaN nanowire junction LEDs often needed the emissions to be recorded with charged-coupled device (CCD) cameras. We suggest that the high EL intensity could be explained by the enhancement of carrier injection in our homojunction nanowire devices. This carrier enhancement in GaN nanowire LEDs could be due to the size effect of the n-GaN nanowires on p-GaN substrates. High electric fields could be induced on p-GaN beneath the n-GaN nanowires due to the small size of n-GaN nanowires and thus the resulting small junction area. This could reduce the depletion width in the p-GaN material and, consequently, increase the tunneling probability[22].

6.3.4 GaN NW half-wave current rectifiers

Fig 6.5 shows the schematic fabrication processes for the fabrication of p-n junction structures. These structures were fabricated by assembling an n-GaN nanowire and a p-Si (100) wafer using ac DEP. The cathode electrodes (Ti/Au=50/100 nm) for contact to n-type GaN nanowires were patterned on an oxidized (500 nm) p-type silicon (100) substrate by a standard photolithography and lift-off process. The anode contacts (Ti/Au=50/100 nm) were defined directly on a p-Si substrate after the native oxide was removed by dipping in dilute hydrofluoric (HF) acid.

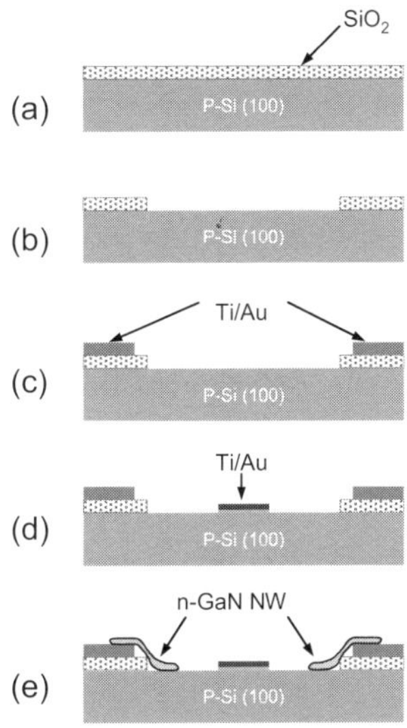

Fig 6.5 A schematic diagram of the process (steps a-e) for fabricating GaN nanowire diode structures, which were formed by dielectrophoresis on a p-Si (100) wafer. The p-n junction structures were assembled by coupling n-GaN nanowires together with p-Si substrates (n-GaN NWs/p-Si) using dielectrophoresis (DEP). (a) Thermal oxidation of the p-Si (100) wafer. (b) Wet-etching of the thermal oxide (SiO_2) with a photo-resist etch mask. (c) Patterning the cathode electrodes (Ti/Au=50/100 nm) on a SiO_2 layer. (d) Patterning the anode electrodes (Ti/Au=50/100 nm) on the p-Si (100) substrate. (e) Dielectrophoresis (DEP) process for the fabrication of n-type GaN nanowires on a p-Si wafer. Annealing was performed at 700-750°C for 30-60 s by RTA (rapid thermal annealing). See Refs. [15 and 18].

To demonstrate the flexibility of these DEP-prepared n-GaN NW/p-Si junction device elements we investigated the operation of n-GaN NW/p-Si-based half-wave rectifiers. Fig 6.6 (a) shows the schematic circuit of a half-wave rectifier. It consisted of a single diode and a load resistor. In general, the half-wave rectifier utilizes alternating half-cycles of the input sinusoid. Using the battery-plus-resistance diode model, the equivalent circuit is shown in Fig 6.6 (b).

From the equivalent circuit, the output voltage (V_{out}) can be written as

$$V_{out} = \frac{R_L}{R_L + R_s} V_{in} - \frac{R_L}{R_L + R_s} V_d \quad (4)$$

Here, V_d is the voltage drop (turn-on voltage) across the p-n junction and R_s is the parasitic resistance. For the calculation, we take the voltage drop (built-in potential of n-GaN/p-Si junction), the load resistance, and the parasitic resistance to be 1.23V, 320 kΩ, and 39 kΩ, respectively. The theoretical total voltage (V_{out}) is then calculated to be 2.907 $V_{p\text{-}p}$ using Eq. (4). In Fig 6.6 (c), the n-GaN NW/p-Si diode structures exhibit a well-defined half-wave current rectifying behavior. As shown in Fig 6.6 (c), when the input voltage (V_{in}) is 10 $V_{p\text{-}p}$, we find that the output voltage (V_{out}) is observed to be 3$V_{p\text{-}p}$. This is clearly in good agreement with the theoretical calculations using Eq. (4).

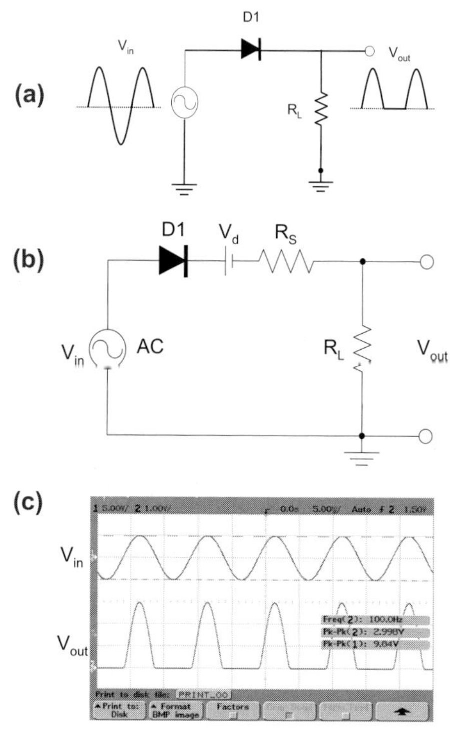

Fig 6.6 (a) The circuit of half-wave rectifier. (b) An equivalent circuit of the half-wave rectifier. (c) Input and output waveforms using GaN nanowire rectifiers which are fabricated by coupling of n-GaN nanowires and p-Si substrates using a DEP process. See Ref. [18].

6.4 Summary

In summary, a simple and effective dielectrophoresis technique with semiconductor GaN nanowires is presented in this chapter. We study the influence of the type of electric field and the applying frequency on the control of GaN nanowires using dielectrophoresis. Stronger ac electric fields force GaN nanowires to align across electrodes with the assembly yield approaching ~80% over the entire electrode array in the chip. This can be simply explained by high dielectrophoresis force in the gap across the electrodes. The yield results indicate that ac dielectrophoresis is much more effective in aligning and manipulating nanowires in the gap than dc electric field. The aligned single or bundle type nanowires are capable of acting as building blocks for fabricating high sensitivity and selectivity chemical sensors as well as high-performance nano-devices. In addition, we demonstrate that GaN NW p-n junction LEDs, which are fabricated by coupling n-GaN nanowires and p-GaN substrates using DEP techniques, show high-brightness UV-blue LED behavior. We also notice that the DEP-prepared GaN nanowire rectifiers (n-GaN NW/p-Si substrates) can be used as half-wave current rectifiers.

Acknowledgements

This work was supported by Korea Research Foundation Grant funded by Korean Government (MOEHRD) (KRF-2005-005-J07501) and (KRF (KRF-2006-331-C00109). We thank Professor H.-J Choi at Yonsei University, Korea and Professor Peidong Yang at the University of California, Berkeley, USA for fruitful discussions.

References

1. Y. Xia, P. Yang, Y. Sun, Y. Wu, B. Mayers, B. Gates,Y. Yin, F. Kim and H. Yan, *One-dimensional nanostructures: synthesis, characterization, and applications*, Adv. Mater. 15, 353 (2003)
2. J. Goldberger, R. He, Y. Zhang, S. K. Lee, H. Yan, H. J. Choi and P. Yang, *Single-crystal gallium nitride nanotubes*, Nature 422, 599 (2003)
3. Y. Huang, X. Duan, Y. Cui and C. M. Lieber, *Gallium nitride nanowire nanodevices*, Nano. Lett. 2, 101 (2002)

4. C. C. Tang, S. Fan, H. Y. Dang, P. Li and L. M. Liu, *Large-scale synthesis of single crystalline gallium nitride nanowires*, Appl. Phys. Lett. 75, 2455 (1999)
5. T. Kuykendall, P. Pauzauskie, S. K. Lee, Y. Zhang, J. Goldberger and P. Yang, *Metalorganic chemical vapor deposition route to GaN nanowires with triangular cross sections*, Nano. Lett. 3, 1063 (2003)
6. L. Zheng, S. Li, J. P. Brody and P. J. Burke, *Manipulating nanoparticles in solution with electrically contacted nanotubes using dielectrophoresis*, Langmuir 20, 8612 (2004)
7. R. Krupke, F. Hennrich, H. B. Weber, M. M. Kappes and H. V. Lohneysen, *Simultaneous deposition of metallic bundles of single-walled carbon nanotubes using ac dielectrophoresis*, Nano. Lett. 3, 1019 (2003)
8. L. Dong, V. Chirayos, J. Bush, J. Jiao, V. M. Dublin, R. V. Chebian, Y. Ono, K. F. Conley Jr. and B. D. Ulrich, *Floating-potential dielectrophoresis-controlled fabrication of single-carbon-nanotube transistors and their electrical properties*, J. Phys. Chem. B 109, 13148 (2005)
9. J. Chung, K. H. Lee and J. Lee, *Nanoscale gap fabrication by carbon nanotube-extracted lithography (CEL)*, Nano. Lett. 3, 1029 (2003)
10. R. Krupke, F. Hennrich, M .M. Kappes and H. V. Lohneysen, *Surface conductance induced dielectrophoresis of semiconducting single-walled carbon nanotubes*, Nano. Lett. 4, 1395 (2004)
11. J. Chung, K. H. Lee, J. Lee and R. S. Ruoff, *Toward large-scale integration of carbon nanotubes*, Langmuir 20, 3011 (2004)
12. D. L. Fan, F. Q. Zhu, R. C. Cammarata and C. L. Chien, *Manipulation of nanowires in suspension by ac electric fields,* Appl. Phys. Lett. 85, 4175 (2004)
13. L. Dong, J. Bush, V. Chirayos, R. Solanki, J. Jiao, Y. Ono, J. F. Conley Jr. and B. D. Ulrich, *Dielectrophoretically controlled fabrication of single-crystal nickel silicide nanowire interconnects,* Nano. Lett. 5, 2112 (2005)
14. X. Duan, Y. Cui, J. Wang and C. M. Lieber, *Indium phosphide nanowires as building blocks for nanoscale electronic and optoelectronic devices,* Nature 409, 66 (2001)
15. T. H. Kim, S. Y. Lee, N. K. Cho, H. K. Seong, H. J. Choi, S. W. Jung and S. K. Lee, *Dielectrophoretic alignment of gallium nitride nanowires (GaN NWs) for use in device applications*, Nanotechnology 17, 3394 (2006)

16. S. Y. Lee, T. H. Kim, D. I. Suh, N. K. Cho, H. K. Seong, S. W. Jung, H. J. Choi and S. K. Lee, *A study of dielectrophoretically aligned gallium nitride nanowires in metal electrodes and their electrical properties*, Chem. Phys. Lett. 427, 107 (2006)

17. S. K. Lee, T. H. Kim, S. Y. Lee, K. C. Choi and P. Yang, *High-brightness gallium nitride nanowire UV-blue light emitting diodes*, Phil. Mag. 87, 2105 (2007)

18. S. Y. Lee, T. H. Kim, E. K. Suh, N. K. Cho, W. K. Seong and S. K. Lee, *Dielectrophoretically aligned GaN nanowire rectifiers*, Appl. Phys. A 87, 739 (2007)

19. Pohl H A, *Dielectrophoresis* (Cambridge University Press: Cambridge, UK, 1978).

20. H. Morgan and N. G. Green, *Dielectrophoretic manipulation of rod-shaped viral particles,* J. Electrostatics 42, 279 (1997)

21. J. Johnson, H. J. Choi, K. P. Knutsen, R. D. Schaller, P. Yang and R. J. Saykally R J, *Single gallium nitride nanowire lasers*, Nature Mater. 1, 106 (2002)

22. B. G. Streetman and S. Banerjee, *Solid state electronic devices* (Prentice Hall Inc.: New Jersey, USA, 2000)

23. http://www.shellchemicals.com

7

DESIGN, FABRICATION, AND APPLICATIONS OF LARGE-AREA WELL-ORDERED DENSE-ARRAY THREE-DIMENSIONAL NANOSTRUCTURES

Chang-Hwan Choi and Chang-Jin "CJ" Kim

A highly effective fabrication process combining interference lithography with deep reactive ion etching is described. This simple technique makes it possible to build well-ordered dense-array of nanostructures with good control of pattern regularity and size over a large area. The unique three-dimensional structures possible with this process can have a variety of applications in electronics and other physical sciences.

7.1 Introduction

Much of nanotechnology as we know it today is based on our ability to create objects of nanometer size scale whether chemically (typically bottom-up) or physically (typically top-down). While any fabrication technique that produces new materials or nanostructures may spur some scientific discoveries that were not possible before, engineering breakthroughs impose much more stringent criteria on the fabrication methods, such as controllability, repeatability, and yield. In some applications, more specifically, one may need nanostructures with good regularity as well as controlled pattern, size, and shape. In some applications, furthermore, nanostructures may not be practical unless they cover a relatively large area.

While numerous nanostructures, made by many different nanoscale patterning and fabrication methods, have been reported[1], we still lack the technology to achieve nanostructure formation with enough regularity and controllability of pattern, size, and shape, to enable expedient and precise scientific studies and engineering applications at the nanoscale. Although serial nano-lithographic techniques such as e-beam, ion-beam, or scanning probe lithography enable direct writing of complex nanoscale patterns, the slow speed of these patterning techniques is not suitable to cover large areas (these techniques typically cover less than 1 mm^2 at a time). While the parallel method of X-ray lithography can pattern a large area, it is too expensive for most applications. On the other hand, soft lithography-based methods, such as nanoimprint lithography, replicate patterns in a parallel fashion but need a master mold which should first be manufactured by e-beam or X-ray lithography. Most non-lithographic methods, such as the use of nano templates constituted by self-assembled nanoparticles (e.g., block copolymers or colloidal nanospheres) or nanoporous membrane (e.g., anodic alumina membrane), lack the regularity some applications demand over a large area. For other non-lithographic methods such as the direct growth of nanostructures (e.g., carbon nanotubes or nanofibers), the controllability and regularity of the pattern size and geometry still remains an issue. Various techniques to fabricate a dense array of nanoscale posts have been developed and evaluated in reference 2.

Currently, interference lithography is considered the most efficient way to make nanoscale periodic patterns over a large area (reportedly up to ~1 m^2) with precise control of regularity and accuracy[3-12]. It uses simple and relatively inexpensive optics to generate uniform interference patterns such as lines and dots on a substrate. It does not use any photomask and has practically unlimited depth of focus. Because of the periodic nature of the patterns created by interference lithography, optical gratings[13,14] or field emitter arrays[15-18] have been fabricated efficiently. However, it should be noted that to transfer the lithographic pattern into the substrate, in particular to etch tall or high-aspect-ratio nanostructures, a thin photoresist (PR) (typically tens of nanometers thick) needs to be replaced by a hard etch mask, such as a metal[3,6,15-17] or an oxide[7,14], before the subsequent deep etching step. The added mask steps make the process more complex and degrade the pattern resolution possible.

Greatly simplifying the process and improving the accuracy of the pattern transfer, we have recently developed a new means of coupling the interference lithography directly with deep reactive ion etching (DRIE)[19]. Noting that DRIE has a very high etch selectivity for silicon over PR (e.g., ~75:1)[20], our approach is to utilize the thin PR layer patterned by the interference lithography directly as an etch mask layer in DRIE, while retaining the traits of both techniques. Well-ordered (post and grate) dense-array (230 nm in pitch) silicon nanostructures with less than 10% deviation in size and shape could be achieved over a large sample area (2 × 2 cm^2). The new combination of DRIE with interference lithography enabled even high-aspect-ratio (higher than 10:1) tall nanostructures (over 1 μm) in one process flow. Commonly used to etch deep trenches with vertical sidewalls in Micro Electro Mechanical Systems (MEMS) fabrication, a Bosch DRIE process (as opposed to a cryogenic process[21]) has rarely been used to construct nanostructures because the well-known effect of sidewall rippling, so-called 'scalloping', is too prominent on the nanoscale. However, recently we have shown that by properly regulating etching parameters in the process recipe, the nanoscopic scalloping problem can be not only controlled but also utilized to realize three-dimensional (3D) nanostructures with sophisticated sidewall profiles[19]. In that work, it was further demonstrated that the tip sharpness can also be controlled by a simple additional process of thermal oxidation and the subsequent removal of the oxide. The well-defined nanostructures over a large area with designable sidewall profiles and tip shapes open new application possibilities in areas beyond electronics and photonics. In this chapter, we will revisit the new nanofabrication technique[19], explaining the design and fabrication of 3D nanostructures followed by their applications.

7.2 Design and Fabrication

7.2.1 Materials and methods

Figure 7.1 shows the overall fabrication process for large-area well-ordered dense-array 3D silicon nanostructures using interference lithography followed by DRIE. A polished silicon substrate (~ 2 × 2 cm^2) is cleaned with a Piranha solution (H_2SO_4:H_2O_2, 3:1 by volume) and dehydrated for 10 minutes at 150°C. The SPR3001 photoresist (Shipley Company, Marlborough, MA) is then spin-coated at 5000 rpm for 1 minute, which gives ~50 nm film thickness. After the spin-coating, a soft-bake is done at 95°C for 1 minute on a hot plate. The substrate is then exposed under an interference lithography setup of a HeCd laser emitting at a wavelength of 325 nm (Fig 7.2). While the pattern periodicity or pitch, p, is tunable by rotating the sample stage (i.e., changing the angle, θ), we specifically set θ equal to 45°, which produces p = 230 nm. Two different regular patterns have been created: nanograte structures using a pattern of parallel lines and nanopost structures using a pattern of a dot array. Such PR dots in a grid array are obtained by two successive exposures with the substrate rotated by 90° in its plane between the exposures. After the exposures, the substrate is developed with MF701 developer (Shipley Company) for 20 seconds, rinsed with de-ionized water, blown dry with N_2 gas, and hard-baked for 1 minute at 110°C on a hot plate.

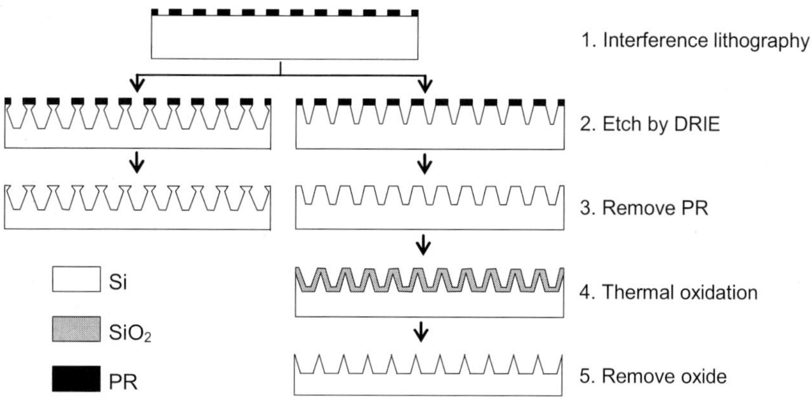

Fig 7.1 Fabrication process of 3D nanostructures with sidewall profile and tip sharpness control[19]. PR pattern created by interference lithography is used as a direct etch mask in DRIE. DRIE etching step involves the design of etching parameters for 3D nanostructure fabrication, e.g., sidewall control for either re-entrant (left column) or positively-tapered (right column) profile. Nanostructures with positively-tapered, smooth sidewall profile can further be sharpened by thermal oxidation and the removal of the oxide.

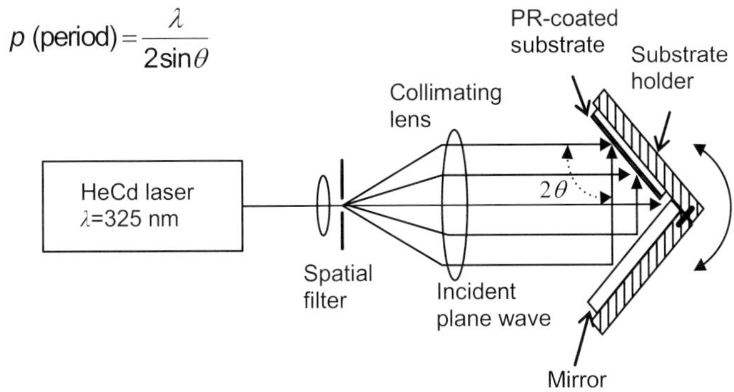

Fig 7.2 Schematic of Lloyds-mirror configuration of laser interference lithography system[19]. A HeCd laser beam with a wavelength of λ=325 nm is expanded and spatially filtered through a pinhole in order to render it coherent. The coherent beam is collimated and aligned toward a mirror and a substrate holder forming a 90° dihedral angle. It inscribes a periodic line pattern of interfered light intensity on a PR-coated substrate. A line pitch p is determined by $p = \lambda/(2\sin\theta)$, where θ is one half of the angle between the directly incoming and the mirror-reflected lights.

The substrate is then etched by DRIE using the patterned PR as an etch mask. For the DRIE etching, we use PlasmaTherm SLR770 ICP etcher (Unaxis Corporation, St. Petersburg, FL). The etching procedures and parameters for the cyclic steps of the Bosch process are shown in Fig 7.3 and summarized in Table 7.1, respectively. One etch cycle consists of two consecutive etch steps of 'Etch A' and 'Etch B' and one 'Deposition' step. Although several parameters in DRIE, such as pressure, RF power, and gas mixture influence the sidewall profile[20-24], the relative duration of etching time (e.g., Etch B time) versus deposition time in the cyclic Bosch process is easier to control with good reproducibility and is thus mostly utilized for the control of scalloping effect and the fabrication of 3D nanostructures. Two means are shown in Fig 7.1 to obtain 3D nanostructures with various sidewall profiles: one with a re-entrant profile on top and the other with a positively-tapered profile. After the DRIE, the remaining PR is removed by O_2 plasma ashing and the sample is cleaned by the Piranha solution. For further modification, such as needle-like sharp-tip nanostructures, the tips of nanostructures with the positive slope are sharpened by the oxidation of silicon and the subsequent removal of silicon dioxide.

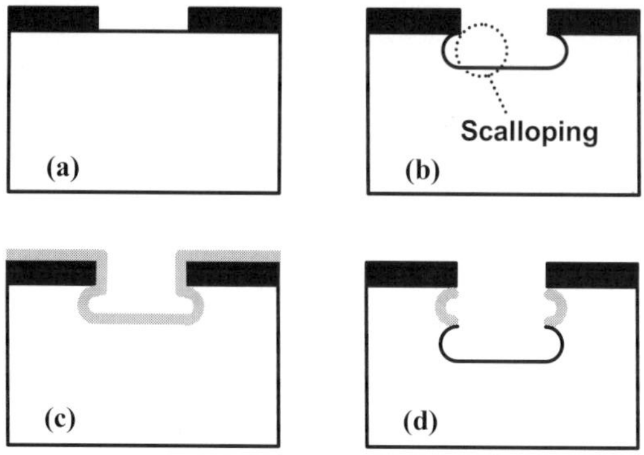

Fig 7.3 Cyclic Bosch process in DRIE[19]. (a) PR opening for Bosch DRIE. (b) Isotropic SF_6 etch with anisotropic bombardment ('Etch A' and 'Etch B' in Table 1). (c) Isotropic polymer formation with C_4F_8 ('Deposition' in Table 1). (d) The SF_6 etch and polymer deposition is repeated according to the total number of etch cycles. Scallops, whose peak-to-valley height is over 50 nm in typical DRIE[20, 22], appear on the walls due to the isotropic nature of the etch. The nanoscale scalloping effect is controlled and utilized for sidewall profile and tip sharpness control for the 3D nanostructure fabrication.

Table 7.1 Parameters in the cyclic Bosch process in DRIE.

Cyclic step [a]	Time (s)	Pressure (mTorr)	C_4F_8 (sccm)	SF_6 (sccm)	Ar (sccm)	RF1 [b] (W)	RF2 [c] (W)
Etch A	2	23	0.5	15	12	5	825
Etch B	**Varied**	23	0.5	30	12	5	825
Deposition	5	23	24	0.5	12	1	825

[a] During the entire series of cyclic steps, the substrate is cooled with helium to keep a constant temperature of ~20°C.

[b] RF1 is the power of reactive ion etch (RIE) generator, capacitively coupled to a wafer chuck to independently bias the substrate.

[c] RF2 is the power of the inductively coupled plasma (ICP), generating very dense plasma near the top of the electrode.

7.2.2 Design and fabrication of "candlestick-like" nanostructures

Although the nano-scalloping effect inherent in the Bosch DRIE is undesirable in typical applications, an intricate sidewall profile for 3D nanostructures can be achieved by planning the degree of the nano-scalloping effect along the sidewall. For example, Fig 7.4 shows sidewall profiles programmed to be re-entrant or "candlestick-like" (i.e., negative right near the top but positive on the rest of the nanoposts below). The degree of the re-entrance was controlled by the scalloping size through the first cycle of the Bosch process. The Etch B time of only the first cycle was increased from 4 seconds to 6, 8, and 10 seconds, while the Etch B time of the subsequent cycles was fixed at 2 seconds. As the Etch B time of the first cycle increased, the scalloping at the beginning (i.e., tip of the nanostructures) became more pronounced and produced a re-entrant (i.e., negative) profile. Below the tip region, the sidewall profile made by the etch cycles using a shorter Etch B time (2 seconds) became smooth and positively-tapered.

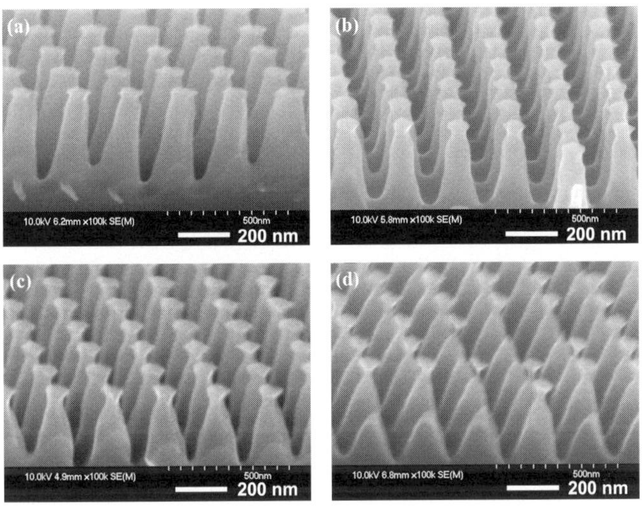

Fig 7.4 Scanning electron microscope (SEM) images of silicon nanopost structures with re-entrant sidewall profiles[19]. Sidewall profile was programmed to change from negative to positive at the tip by controlling the scalloping size (i.e., Etch B time) of the first cycle of the Bosch process. (a) 1 cycle with Etch B = 4 s, followed by 10 cycles with Etch B = 2 s. (b) 1 cycle with Etch B = 6 s, followed by 8 cycles with Etch B = 2 s. (c) 1 cycle with Etch B = 8 s, followed by 8 cycles with Etch B = 2 s. (d) 1 cycle with Etch B = 10 s, followed by 8 cycles with Etch B = 2 s.

7.2.3 Design and fabrication of "hierarchical" nano structures

The local sidewall profile (i.e., the local formation of nano-scalloping along the sidewall) and the overall sidewall slope (tapering) are interactively affected not only by the relative duration of the etch time (i.e., Etch B time) against the deposition time, but also by the total number of etch cycles[19]. Although the regulation of Etch B time makes the creation of 3D nanostructures practicable, it is not easy to independently control the local (i.e., at a given depth) sidewall profile and the overall sidewall slope simultaneously. In addition to the Etch B time, the modulation of the other parameters such as RF power, pressure, and gas mixture is also desired for the independent control of the local sidewall profile and the overall sidewall slope. For example, Fig 7.5 shows nanopost structures whose overall sidewall slope was designed to be all positive but whose local sidewall profiles vary independently. The different degrees of nano-scalloping formation along the sidewall, while maintaining the overall positive-slope of the sidewall, was enabled by modulating several etch parameters concurrently. This result, as clearly shown in Figs 7.5(c) and 7.5(d), suggests that independent control of the

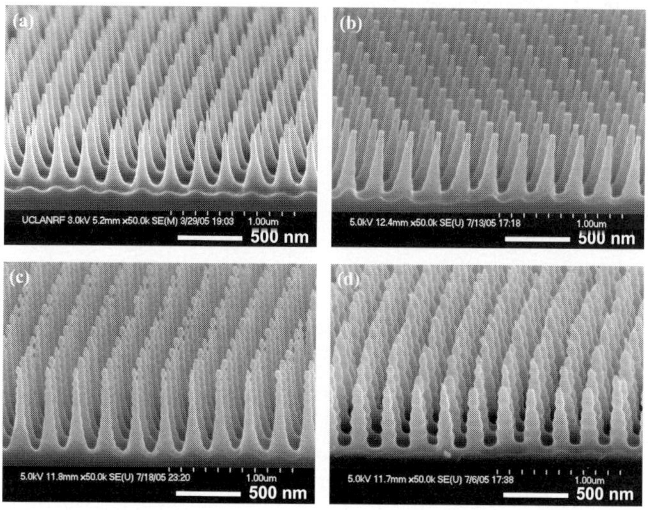

Fig 7.5 SEM images of silicon nanopost structures with the same positively-tapered sidewalls but with different roughness[19]. Sidewall tapering (slope) was controlled to be all positive, while different degrees of nano-scalloping effects were imposed on the sidewalls by modulating the other etch parameters set in Table 7.1. (a) 18 cycles with Etch B = 1 s, while the other parameters are the same as shown in Table 1; used as a reference. (b) 10 cycles with Etch B = 2 s, deposition = 4 s, pressure = 20 mTorr, and RF1 (for Etch A and Etch B) = 9 W. (c) 15 cycles with Etch B = 1 s, deposition = 3 s, pressure = 20 mTorr, and RF1 (for Etch A and Etch B) = 9 W. (d) 8 cycles with Etch B = 3 s, deposition = 5 s, pressure = 20 mTorr, gas flow rates = 15% increase, and RF1 (for Etch A and Etch B) = 9 W.

lower-order (smaller) nano-scalloping formation along the surface of the higher-order (larger) nanostructure will enable a multidimensional "hierarchical" nanostructure. This result further supports the idea that the well-programmed nano-scalloping effect by modulating the etch parameters in Bosch DRIE can be a simple but useful tool to tailor the sidewall profile of 3D micro and nanostructures.

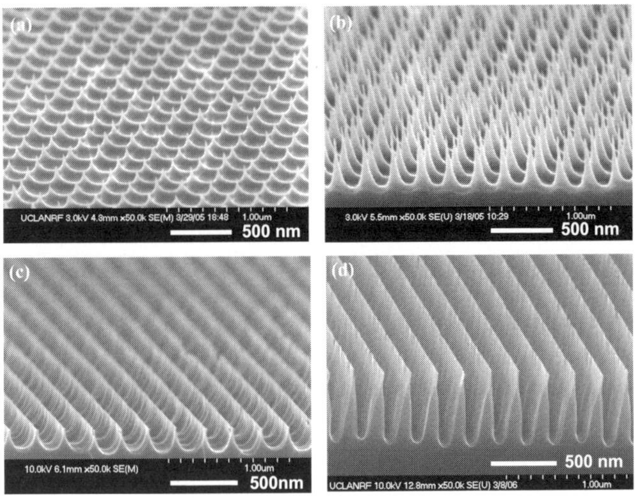

Fig 7.6 SEM images of silicon nanostructures with sharp tips[19]. Tips of nanostructures with positively-tapered, smooth sidewall profiles were sharpened by thermal oxidation followed by oxide etching, measuring less than 5 nm in tip apex radius of curvature. (a)-(b) Nanopost structures with needle-like sharp tips (a: 114 ± 11 nm in height and 11 ± 1° in cone angle, b: 460 ± 17 nm in height and 5 ± 1° in cone angle). (c)-(d) Nanograte structures with blade-like sharp tips (c: 180 ± 10 nm in height and 12 ± 1° in cone angle, d: 600 ± 18 nm in height and 4 ± 1° in cone angle).

7.2.4 Design and fabrication of "sharp-tip" nanostructures

As proposed in Fig 7.1, sharp-tip nanostructures can be developed from nanostructures with a positively-tapered, smooth sidewall profile. For example, Fig 7.6 shows sharp-tip nanostructures of varying heights. While the size (i.e., height or aspect ratio) of the sharp-tip nanostructures is initially determined in DRIE, it can also be modulated further in the timed oxidation step. The cone angle of the sharp-tip nanostructures is also designable with the sidewall slope control in the DRIE step, while the subsequent thermal oxidation and wet oxide etch do not change the cone angle significantly since they are isotropic processes. The tip apex radius of curvature of the sharp-tip nanostructures shown in Fig 7.6 is less than 5 nm, regardless of the size and shape; all were measured by the image analysis of cross-sectional SEM images.

7.3 Applications

To create 3D structures with conventional fabrication techniques, multiple lithography steps with precise alignment or multi-layer resists (or multi-step post processes) within a single lithography step would be required. In this regard, the simple but efficient fabrication method of 3D nanostructures presented here, directly achievable in just one process flow in Bosch DRIE, is much more suitable for various practical applications. For example, 3D nanostructures are desirable in several applications such as T-gates for microwave transistors, wave modulators for nano-optics, and various nanoelectromechanical systems (NEMS). For a given void fraction under the surface, nanostructures with a re-entrant profile provide less open and more flat area on the surface than simple profiles do. One use of such a re-entrant profile would be to produce monolithic nanochannels by sealing the top of the nanograte structures with a thin-film coating. The well-ordered sharp-tip nanopost structures covering a large pattern area are commonly of interest in such electronic applications as field emission structures. The simple way outlined here to fabricate sharp-tip nanostructures will also facilitate the design and fabrication of high-aspect-ratio scanning probe tips. Well-ordered densely-populated nanostructures over a large sample area could also be useful for non-electronic applications as described below.

7.3.1 Low-friction superhydrophobic surface

One novel non-electronic application of sharp-tip nanostructures is to reduce friction of liquid flows. After a hydrophobic surface treatment, sharp-tip nanostructures make the surface superhydrophobic because the liquid is levitated over the air layers sustained among the non-wetting nanostructures (Fig 7.7). In this way, the liquid has minimal contact with the solid surface (i.e., contacting only the tips of the nanostructures), and is expected to flow with significantly reduced skin friction (i.e., significantly increased slip flow). Although the basic concept has been known from nature[25,26] and tested[27-29] for years, there has not been a deliberate effort to *design* and *fabricate* a surface to produce a meaningful reduction in friction under practical conditions (e.g., highly pressurized flows frequently encountered in engineering practice). Three main new features are critical for the design and fabrication of low-friction superhydrophobic surface[30-32]. First, the nanostructures should be populated with *submicron density*, which keeps the surface dry even under pressurized liquid. Similar surface structures with micron scale density would be filled with liquid under a nominal pressure (e.g., > 1 atm) and lose its functionality. Second, the nanostructures should be *tall and slender*, providing a thick air layer below the liquid. Without enough air to lubricate, the flow friction is not reduced enough to be meaningful. Third, the *sharpness* of the nanostructures should minimize the liquid-solid contact. The well-controlled, well-ordered sharp-tip nanostructures have a dense pitch on the nanoscale enabling a detailed study of the effect of nanostructure geometries on superhydrophobicity. Such structures also show great promise in practical flow applications by tolerating highly pressurized flows without losing superhydrophobicity. Although the superhydrophobic surface is commonly thought of as conducive to help reduce pressure drops for microchannel flows, its

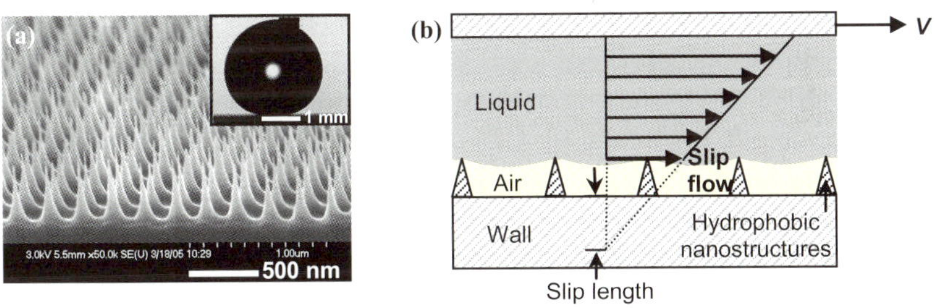

Fig 7.7 Application of sharp-tip nanostructures to low-friction superhydrophobic surface. (a) SEM images of a sharp-tip nanopost surface[19]. The inset shows the apparent contact angle of water droplet on the surface (~180°) after a hydrophobic coating with Teflon (~10 nm). (b) Concept of a low-friction surface for liquid flow (e.g., in Couette flow)[31]. Liquid sits on the hydrophobic sharp-tip nanostructures by surface tension. The majority of the liquid boundary is with air, where shear stress is much smaller. The hydrophobic nanostructures need to be relatively tall and populated on the submicron scale to sustain the liquid levitation and to produce effectively large slip flow over air under realistic conditions (i.e., pressure ranging up to 1 atm).

utility is much wider. For instance, the large slip flow at walls can help flatten the velocity profiles within microchannels, which could then be utilized to reduce the dispersion in microfluidic separation systems.

7.3.2 Novel substrate for nanobioscience and nanobioengineering

The well-regulated nano-topographical properties of our nanostructures will also provide a unique testbed enabling a number of detailed studies in nanobioscience and nanobioengineering. For example, cell-matrix adhesion *in vivo* is a 3D phenomenon that differs from the adhesion on two-dimensional (2D) substrates *in vitro*[33]. Within the extracellular matrix, cells interact with nanoscale topographical projections and depressions that vary in composition, size and periodicity[34]. The matrix topography is important for proper adhesion and the activation of desired intracellular pathways, affecting cell behavior in terms of morphology, cytoskeletal arrangement, migration, proliferation, surface antigen display, and gene expression[35,36]. Although several cell behaviors over various surface topographies had been studied with micro- and nanostructured surfaces[35,36], the inability to independently control nanodimensionality and nanoperiodicty in the nanoscale range has to date precluded a systematic study of the 3D effects of nanoscale features on cell behaviors. As shown in Fig 7.8, our well-defined nanostructured surfaces provide a unique opportunity to elucidate many aspects of the nanobiology of the cell, including the 3D effect of the surface nanotopography on cell behaviors, whose understanding can further be utilized for cell and tissue engineering applications[37].

For example, the lateral tip size of our sharp-tip nanostructures (less than 10 nm in tip apex radius of curvature), comparable to that of a single integrin molecule (8-12 nm) in a cell membrane, can provide unique capability to examine integrin activation and focal adhesion on 3D nanotopographies, which is essential for adhesion-mediated signaling. Our sharp-tip nanotopographies capable of excellent control of nanodimensionality and nanoperiodicity will also enable the investigation of relative contributions and interactions between nanotopographical three-dimensionality and periodicity on integrin clustering and activation. Fewer cell populations with the retardation of cell growth observed on the sharp-tip tall nanopost structures suggest that the needle-like sharp-tip nanostructures should be useful for a biological low adhesive surface, i.e., anti-adhesion or anti-fouling surface. The control of bio-adhesion or bio-fouling only through control of the surface nanotopographies, as opposed to chemical modification, will provide many advantages in the design of biomaterials such as biomedical implant surfaces. Another advantage of the needle-like sharp-tip nanostructures is that they can be used as drug delivery or biochemical manipulation systems. The needle-like sharp-tip nanoposts can be inserted into viable cells, by pressing for instance, and transport chemicals pretreated on the sharp-tip surfaces across cell membranes. Furthermore, they can be utilized as an intracellular interface for monitoring and controlling subcellular and molecular phenomena in the way of *in vivo* biosensors and actuators. The significant alignment with elongation of the nanograte topographies (see Fig 7.8b) also suggests the possibility of controlling the cells' orientation or structure by using directional nanostructures, which may be desirable in tissue engineering applications. For instance, nanotopography may be exploited to create cell sheets with specified cell-alignment patterns, and then layers of nanoengineered cell sheets can be stacked to create 3D tissue constructs for tissue regeneration applications.

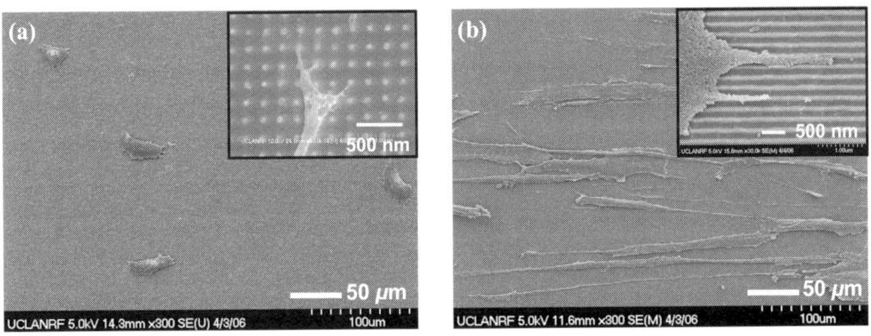

Fig 7.8 SEM images of human foreskin fibroblast cells cultured on sharp-tip nanostructure surfaces (a: nanopost, b: nanograte)[37]. Each inset is a magnified image of the filopodia of cells interacting with the surface[19]. Human foreskin fibroblasts exhibited significantly smaller cell size and lower proliferation on nanoposts, and enhanced elongation with alignment on nanogrates. These phenomena became more pronounced as the nano-topographical three-dimensionality (structural height) increased. The nanopost and nanograte architectures provided the distinct contact guidance for both filopodia extension and the formation of adhesion molecules complex, which was believed to lead to the unique cell behaviors observed.

7.4 Conclusion

A simple and effective fabrication method for building well-ordered dense-array of nanostructures with good control of pattern regularity and size over a large area was achieved by combining interference lithography with DRIE in one process flow. The nanoscale scalloping effect in Bosch DRIE was controlled and utilized for creating novel 3D nanostructures with various sidewall profiles and tip shapes. Affordable surfaces with well-controlled nanostructures over a large area open new applications not only in electronics but also in the physical world through their unique properties originating from their nanoscale geometry. We envision that various 3D nanostructures, from not only silicon but also other materials such as metal, glass, and polymers, can be designed and fabricated for further novel applications; incorporating the current techniques in the process.

Acknowledgements

This research has been funded by the NSF (National Science Foundation) NIRT (Nanotechnology Interdisciplinary Research Team) Grant 0103562 and California NanoSystems Institute. The authors thank Nanotech at the University of California at Santa Barbara for their help with the use of the interference lithography system. We also appreciate Dr. Joonwon Kim for his initial help in nanofabrication, Profs. Chih-Ming Ho and Pirouz Kavehpour for their advice on slip flow measurements, and Profs. Benjamin Wu, James Dunn, and Dr. Sepideh Hagvall for their consultation on cell studies and tissue engineering applications.

References

1. J. I. Martin, J. Nogues, K. Liu, J. L. Vicent and I. K. Schuller, *Ordered magnetic nanostructures: fabrication and properties*, J. Mag. Mag. Mat. 256, 449 (2003)
2. J. Kim, Movement of liquid metal and aqueous solution on micro- and nano-engineered non-wetting surfaces, Ph.D. dissertation, Mechanical and Aerospace Engineering Department, University of California, Los Angeles, 2003.
3. E. H. Anderson, C. M. Horwitz and H. I. Smith, *Holographic lithography with thick photoresist*, Appl. Phys. Lett. 43, 874 (1983)

4. S. H. Zaidi and S. R. J. Brueck, *High aspect-ratio holographic photoresist gratings*, Appl. Opt. 27, 2999 (1988)
5. S. H. Zaidi and S. R. J. Brueck, *Multiple-exposure interferometric lithography*, J. Vac. Sci. Technol. B 11, 658 (1993)
6. M. L. Schattenburg, R. J. Aucoin and R. C. Fleming, *Optically matched trilevel resist process for nanostructure fabrication*, J. Vac. Sci. Technol. B 13, 3007 (1995)
7. T. A. Savas, M. L. Schattenburg, J. M. Carter and H. I. Smith, *Large-area achromatic interferometric lithography for 100 nm period gratings and grids*, J. Vac. Sci. Technol. B 14, 4167 (1996)
8. A. Fernandez, J. Y. Decker, S. M. Herman, D. W. Phillion, D. W. Sweeney and M. D. Perry, *Methods for fabricating arrays of holes using interference lithography*, J. Vac. Sci. Technol. B 15, 2439 (1997)
9. G. Gigli, R. Rinaldi, C. Turco, P. Visconti, R. Cingolani and F. Cacialli, *Holographic nanopatterning of the organic semiconductor poly(p-phenylene vinylene)*, Appl. Phys. Lett. 73, 3926 (1998)
10. H. H. Solak, D. He, W. Li, S. Singh-Gasson, F. Cerrina, B. H. Sohn, X. M. Yang and P. Nealey, *Exposure of 38 nm period grating patterns with extreme ultraviolet interferometric lithography*, Appl. Phys. Lett. 75, 2328 (1999)
11. M. L. Schattenburg, C. Chen, P. N. Everett, J. Ferrera, P. Konkola and H. I. Smith, *Sub-100 nm metrology using interferometrically produced fiducials*, J. Vac. Sci. Technol. B 17, 2692 (1999)
12. H. I. Smith, *Low cost nanolithography with nanoaccuracy*, Physica E 11, 104 (2001)
13. M. L. Schattenburg, C. R. Canizares, D. Dewey, K. A. Flanagan, M. A. Hamnett, A. M. Levine, K. S. K. Lum, R. Manikkalingam, T. H. Markert and H. I. Smith, *Transmission grating spectroscopy and the advanced X-ray astrophysics facility*, Opt. Eng. 30, 1590 (1991)
14. S. H. Zaidi, A.-H. Chu and S. R. J. Brueck, *Optical properties of nanoscale, one-dimensional silicon grating structures*, J. Appl. Phys. 80, 6997 (1996)
15. C. O. Bozler, C. T. Harris, S. Rabe, D. D. Rathman, M. A. Hollis and H. I. Smith, *Arrays of gated field-emitter cones having 0.32 μm tip-to-tip spacing*, J. Vac. Sci. Technol. B 12, 629 (1994)
16. J. P. Spallas, A. M. Hawryluk and D. R. Kania, *Field emitter array mask patterning using laser interference lithography*, J. Vac. Sci. Technol. B 13, 1973 (1995)

17. J. P. Spallas, R. D. Boyd, J. A. Britten, A. Fernandez, A. M. Hawryluk, M. D. Perry and D. R. Kania, *Fabrication of sub-0.5 μm diameter cobalt dots on silicon substrates and photoresist pedestals on 50 cm x 50 cm glass substrates using laser interference lithography*, J. Vac. Sci. Technol. B 14, 2005 (1996).
18. X. Chen, S. H. Zaidi, S. R. J. Brueck and D. J. Devine, *Interferometric lithography of sub-micrometer sparse hole arrays for field-emission display applications*, J. Vac. Sci. Technol. B 14, 3339 (1996)
19. C.-H. Choi and C.-J. Kim, *Fabrication of Dense Array of Tall Nanostructures over a Large Sample Area with Sidewall Profile and Tip Sharpness Control*, Nanotechnology 17, 5326 (2006)
20. A. A. Ayon, R. A. Braff, C. C. Lin, H. H. Sawin and M. A. Schmidt, *Characterization of a time multiplexed inductively coupled plasma etcher*, J. Electrochem. Soc. 146, 339 (1999)
21. M. J. Walker, *Comparison of Bosch and cryogenic processes for patterning high aspect ratio features in silicon*, Proc. SPIE 4407, 89 (2001)
22. A. A. Ayon, R. A. Braff, R. Bayt, H. H. Sawin and M. A. Schmidt, *Influence of coil power on the etching characteristics in a high density plasma etcher*, J. Electrochem. Soc. 146, 2730 (1999)
23. M. J. de Boer, J. G. E. Gardeniers, H. V. Jansen, E. Smulders, M.-J. Glide, G. Roelofs, J. N. Sasserath and M. Elwenspoek, *Guidelines for etching silicon MEMS structures using fluorine high-density plasma at cryogenic temperatures*, J. MEMS 11, 385 (2002)
24. C. Chang, Y.-F. Wang, Y. Kanamori, J.-J. Shih, Y. Kawai, C.-K. Lee, K.-C. Wu and M. Esashi, *Etching submicrometer trenches by using the Bosch process and its application to the fabrication of antireflection structures*, J. Micromech. Microeng. 15, 580 (2005)
25. T. Wagner, C. Neinhuis and W. Barthlott, *Wettability and contaminability of insect wings as a function of their surface sculptures*, Acta Zoologica 77, 213 (1996)
26. W. Barthlott and C. Neinhuis, *Purity of the sacred lotus, or escape from contamination in biological surfaces*, Planta 202, 1 (1997)
27. K. Watanabe, Y. Udagawa and H. Udagawa, *Drag reduction of Newtonian fluid in a circular pipe with highly water-repellent wall*, J. Fluid Mech. 381, 225 (1999)

28. S. Gogte, P. Vorobieff, R. Truesdell, A. Mammoli, F. V. Swol, P. Shah and C. J. Brinker, *Effective slip on textured superhydrophobic surfaces*, Phys. Fluids 17, 051701 (2005)
29. J. Ou, B. Perot and J. P. Rothstein, *Laminar drag reduction in microchannels using ultrahydrophobic surfaces*, Phys. Fluids 16, 4635 (2004)
30. J. Kim and C.-J. Kim, *Nanostructured surfaces for dramatic reduction of flow resistance in droplet-based microfluidics*, in Proc. 15th Int. Conf. Micro Electro Mechanical Systems (Las Vegas, Nevada) (Piscataway: IEEE), 479 (2002)
31. C.-H. Choi and C.-J. Kim, *Large slip of aqueous liquid flow over a nanoengineered superhydrophobic surface*, Phys. Rev. Lett. 96, 066001 (2006)
32. C.-H. Choi, U. Ulmanella, J. Kim, C.-M. Ho and C.-J. Kim, *Effective slip and friction reduction in nanograted superhydrophobic microchannels*, Phys. Fluids 18, 087105 (2006)
33. E. Cukierman, R. Rankov, D. R. Stevens and K. M. Yamada, *Taking cell-matrix adhesions to the third dimension*, Science 294, 1708 (2001)
34. G. A. Abrams, S. L. Goodman, P. F. Nealey, M. Franco and C. J. Murphy, *Nanoscale topography of the basement membrane underlying the corneal epithelium of the rhesus macaque*, Cell Tissue Res. 299, 39 (2000)
35. R. G. Flemming, C. J. Murphy, G. A. Abrams, S. L. Goodman and P. F. Nealey, *Effects of synthetic micro- and nano-structured surfaces on cell behavior*, Biomaterials 20, 573 (1999)
36. A. Curtis and C. Wilkinson, *Nanotechniques and approaches in biotechnology*, Trends Biotechnol. 19, 97 (2001)
37. C.-H. Choi, S. H. Hagvall, B. M. Wu, J. C. Y. Dunn, R. E. Beygui and C.-J. Kim, *Cell interaction with three-dimensional sharp-tip nanotopography*, Biomaterials 28, 1672 (2007)

8

UV-NIL STAMP FABRICATION TECHNIQUES WITH DIAMOND-LIKE CARBON FILM

Jun-ho Jeong, Ali Ozhan Altun and Ki-don Kim

The fabrication of nanometre-scale Diamond-Like Carbon (DLC) stamps for ultraviolet Nano Imprint Lithography (NIL) is described in this chapter. Stamp fabrication involves DLC coating and Focused Ion Beam (FIB) lithography. The fabrication of 3D patterns with multilayered stamps is demonstrated. Also described are Fluorine-doped Diamond-Like Carbon (F-DLC) stamps with sub-100-nm line patterns, fabricated with a direct etching method. These stamps were successfully used for ultraviolet nanoimprint lithography without the use of any anti-adhesion layer coating.

8.1 Introduction

Nanoimprint lithography (NIL)[1-3] is one of the most promising new technologies for fabricating patterns with resolutions of less than 10 nm because it allows a higher throughput and lower cost than conventional photolithography. Ultraviolet (UV)-NIL[2-3], which is performed at low pressures and room temperature, is particularly advantageous compared to thermal-type NIL. In UV-NIL, a nanopatterned UV-transparent stamp is pressed onto a dispensed resin or spin-coated resin on a substrate, and then the stamp is exposed to UV light from above to cure the resin. After 10–20 seconds, the stamp is separated from the patterned layer on the substrate. As elevated temperature or high pressure is not needed, this method is both sufficiently rapid and prevents stress on the stamp or substrate. Finally, an anisotropic etching is used to transfer the patterns onto the substrate.

Finding easy and cheap methods for the fabrication of the stamp using proper materials is crucial in UV-NIL. The stamp used in UV-NIL must possess a high UV-transmittance in order to transmit the UV light sufficiently to cure the resin in an acceptable amount of time. In addition, the surface energy of the stamp must be sufficiently low to ensure a fine replication of the pattern when it is separated from the cured resin. A low stamp surface energy is crucial, especially when patterns with a high aspect ratio are imprinted. The surface energy and contact angle are inversely proportional quantities, but it is easier to measure the contact angle. The contact angle of stamps used in UV-NIL generally ranges from 80° to 100°. In addition, the mechanical hardness of a stamp is also important if it is to be used for mass production.

Because UV light must penetrate the stamp, it is constructed from quartz or glass, and the stamping features are fabricated using reactive ionic etching to achieve the desired nanoscale patterns. After the process of electron beam lithography, reactive ionic etching is used to etch patterns onto the hard mask layer of the quartz or glass substrate. This patterned hard mask layer is used to etch the substrate. The general procedure for nanoscale stamp fabrication as explained above is very expensive and time-consuming. Since glass and quartz have a high surface energy, an anti-adhesion layer must be coated on the stamp. The anti-adhesion layer has a certain lifetime so the stamp must be recoated occasionally in a mass production line.

Compared to quartz or glass, diamond-like carbon (DLC) has advantages such as lower surface energy and higher hardness. A stamp made from DLC, which has a low surface energy, does not require the use of an anti-adhesion layer and high hardness is a desired property of stamps in mass production environments. A DLC layer can also be produced by a wide range of deposition methods, such as ion beam deposition, magnetron sputtering, and ion beam sputtering. In this chapter, we present several different techniques related to DLC stamp fabrication for UV-NIL.

8.2 Fabrication of Diamond-like Carbon Stamps for Ultraviolet Nanoimprint Lithography

8.2.1 Stamp fabrication technique using DLC coating and focused ion beam lithography

The UV transparency of DLC coated substrate is highly dependent on DLC coating thickness. In the case of coating thickness of 100 nm, UV transparency values were measured to be 5 to 18% in the 350 to 450 nm wavelength range using a UV-visible spectrophotometer as a quartz wafer was used as substrate. In the case of coating thickness of 200 nm, the UV transparency values decreased to the range of 2 to 8%. We typically used the DLC layer with a thickness of 100 nm for the fabrication of a DLC stamp because of the high UV transparency.

We fabricated a 2-D stamp by using the DLC coating process and then FIB lithography. To deposit the DLC layers on top of the quartz substrates of the stamp, we used the process, namely radio frequency plasma-enhanced chemical vapor deposition, at a frequency of 13.56 MHz. The bias voltage for the deposition was 400 V, and each 100 nm thick deposition was conducted for 15 min. Methane gas (CH_4, 99.9999%) was used as a precursor. All the substrates were plasma-cleaned with Argon before the deposition to increase the surface energy and to improve the adhesiveness of the DLC. The base pressure was from 3 Torr to 10 Torr, and the working pressure was maintained at 10 mTorr. The contact angle of the DLC layer was between 60° and 70°.

We used FIB lithography to fabricate lines on 100 nm thick DLC layers coated on quartz substrates: some of the lines were 70 nm deep, 100 nm wide and had a space of 50 nm; the other lines were 70 nm deep, 150 nm wide and had a space of 50 nm. For the FIB patterning, we used a fast scanning electron microscope (FSEM)/FIB dual beam system (FEI Nova200 Nanolab). We also used a focused Ga+ beam with the energy of 30 keV for patterning at beam currents ranging from 10 pA to 50 pA. Figures 8.1(a) to 8.1(b) show the scanning electron microscope images of patterns that formed on a DLC stamp. The wafers were printed on a UV-NIL setup using the DLC stamp and a photocurable resin[5] whose principal component is tripropylene glycol diacylate (TPGDA). Figures

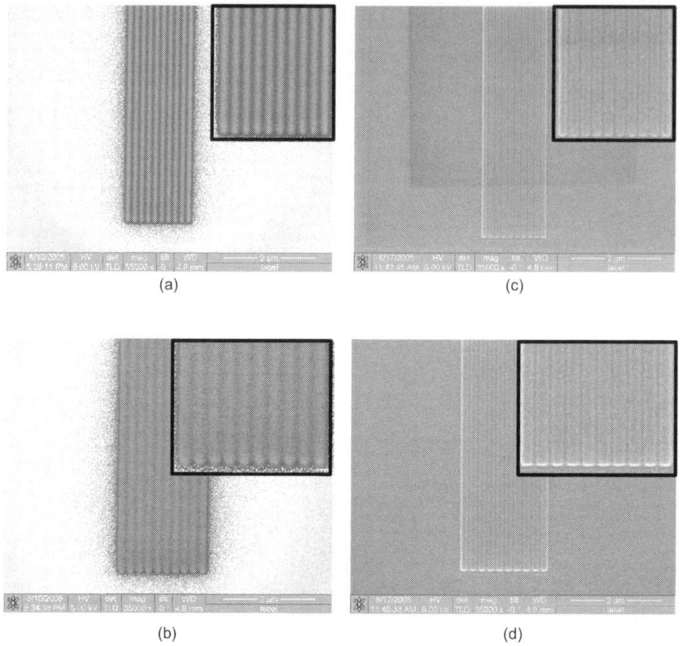

Fig. 8.1 SEM images of DLC stamp with (a) 70 nm deep lines with a width of 100 nm and a space of 50 nm and (b) 70 nm deep lines with a width of 150 nm and a space of 50 nm by fabricated using DLC coating process and FIB lithography. SEM images of corresponding imprinted structures (c-d). (From J. of Nanoscience and Nanotechnology, 6, 1, 2006 Ref: 4.)

8.1(c) and 8.1(d) show the corresponding imprinted features. The features were imprinted at an exposure dose of 14.4 mW/cm^2, an exposure time of 60 s, and an imprint pressure of 930 mbar. The critical dimension (CD) deviations between the features on the DLC stamp and the imprinted features were measured to be a few percent. The shrinkage of the photocurable resin during curing caused the deviations.

8.2.2 Stamp fabrication technique using water-soluble polymer and DLC coating

Recently, several polymer materials have been applied in most types of soft lithography as alternative low-cost stamps, including polydimethylsiloxane (PDMS), polyurethane (PU), and amorphous fluoropolymer. These polymer replica molds of quartz stamps have an obvious advantage in that they can preserve the expensive original master.

However, a one-time replicated stamp from a master stamp has an opposite surface topology to the master stamp. To attain the same stamp surface topology, a replicated stamp must be replicated yet a second time. When PDMS is used as the stamp material in both the first and second replications, it is difficult to separate the two stamps. Although other polymers can be used instead of PDMS as the stamp material in the first replication, the original features are modified by the double effect of polymer shrinking through polymer solidification. Even if these drawbacks could be overcome, transparent polymers such as PDMS are not appropriate materials for the second replication. PDMS has lower mechanical properties (e.g., a low Young's modulus of 1.8 MPa) than harder stamp materials used for imprinting, which has a negative effect on minimizing distortion and maximizing life cycle.

We propose a fabrication process for nanoscale stamps using the UV-NIL process. The stamp's first replication uses a well-known water-soluble polymer, polyvinyl alcohol (PVA). A diamond-like carbon (DLC) is then coated onto the sacrificial PVA replica. Many studies have reported the application of PVA to nanoscale patterning. Spin-cast PVA films have been used to create templates for imprint lithography and molecular transfer lithography[6–9]. Replicated PVA templates have been used to fabricate a polymer nanonozzle array by dissolving the sacrificial template[10].

In this study, we applied a two-step replication to produce an UV-NIL stamp. Figure 8.2 presents the schematics of the replication process. In principle, the master and patterns are without limits. First, the master was coated with an anti-adhesion layer for easy release, after which an aqueous solution of a water-soluble polymer was cast onto the master mold. After drying, the sacrificial template was peeled off and attached to a flat substrate (i.e., a silicon wafer). The DLC was deposited onto this sacrificial template, and then a thick glass plate was affixed to the DLC film for reinforcement using a transparent adhesive. After the adhesive was cured, the reinforced DLC stamp was released by dissolving the sacrificial template in water. Because the sacrificial template was composed of a water-soluble polymer, it therefore allowed fast, easy, and safe removal, unlike previous methods involving chemical etching or heating. This process prevents challenges associated with shrinking, which commonly occur during the conventional replication process.

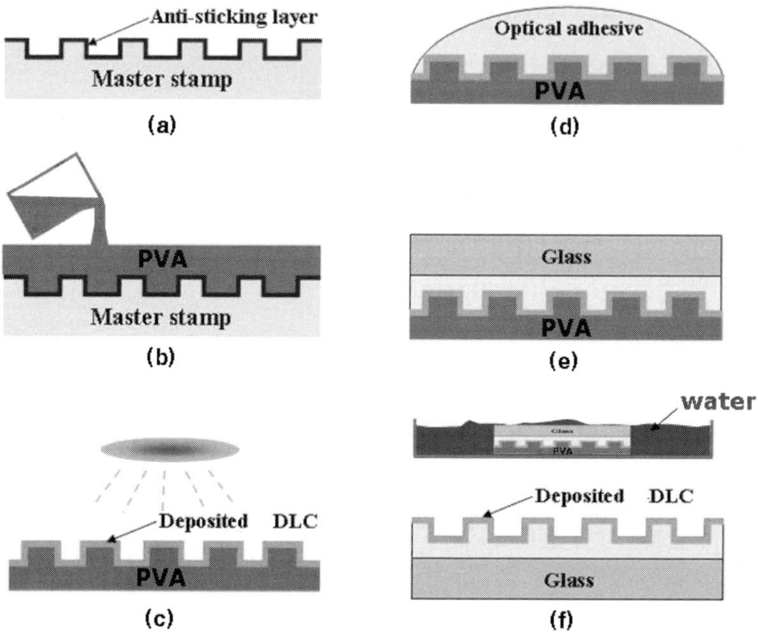

Fig. 8.2 Schematics for the replication of a master stamp using diamond-like carbon (DLC) deposition and polyvinyl alcohol (PVA) molding: (a) surface treatment of the original stamp; (b) pouring PVA onto the original stamp; (c) DLC deposition; (d) pouring the optical adhesive; (e) placing the glass on the adhesive; and (f) dissolving PVA from the adhesive.

The experimental conditions can be summarized as follows. A silicon master nanoscale stamp fabricated using electron beam lithography followed by directional etching was treated by liquid phase deposition of trichloro (1H, 1H, 2H-perfluorooctyl) silane (97%; Aldrich, Milwaukee, WI, USA) for 10 min. The stamp was rinsed with ethanol and acetone following the anti-adhesion treatment. A self-assembled monolayer (SAM) composed of $-CF_3$ formed with this treatment, decreasing surface energy and increasing the contact angle. Most sacrificial templates are composed of PVA, a water-soluble polymer. A 5% PVA (average MW 70,000; Aldrich) aqueous solution was poured over the Si master placed in a Petri dish, after which it was dried overnight at room temperature. After drying, a PVA sacrificial template with a thickness of about 500 μm was peeled off and cut into rectangular slabs while retaining a margin. It was then attached to a 4-inch silicon wafer. Prior to depositing the DLC, pattern-free regions were protected with cellophane adhesive tape. Radio frequency plasma-enhanced chemical vapor was deposited at a frequency of 13.56 MHz to form the DLC layer on top of the PVA sacrificial template. The deposition used a 400-volt bias voltage, and each 30-nm-thick deposition was conducted for 5 min. The UV transparency of a DLC-coated substrate depends highly on the thickness of the DLC. The 100-nm-thick DLC coated glass wafer had UV transparency values of 5 to 18% for wavelengths between 350 and 450 nm. However, DLC coatings less than 10 nm had an UV transparency of approximately 80% with good anti-adhesion characteristics. To strengthen the DLC film, a DLC coating with a thickness of 30 nm was placed on the sacrificial template. Methane gas (CH_4, 99.9999%) was used as a precursor. The base pressure ranged from 3 to 10 Torr, and working pressure was maintained at 10 mTorr. After the protective tape was removed, an UV-curable adhesive (Norland Optical Adhesive 65; Norland Products, Inc., New Brunswick, NJ, USA) was poured onto the DLC-coated PVA sacrificial template. A glass slide was affixed, and it was exposed to UV. After adhesive curing, the DLC layer and glass slide were detached from the Si wafer by dissolving the sacrificial PVA template in hot water. PVA regions not coated with DLC were released very rapidly.

Figures 8.3(a) and 8.3(b) present scanning electron microscopy (SEM) images of the Si master stamp and its atomic force microscope (AFM) image. The checkerboard features are 1μ in width, 1μ in height, and 150 nm in depth. The sacrificial PVA template was replicated from the master stamp (see Figs. 8.2(c) and 8.2(d)). PVA exhibits surfactant characteristics due to the presence of several hydroxyl groups. The solution easily moistened the hydrophobic surface and completely filled the master stamp without air entrapment. Figures 8.3(e) and 8.3(f) present the deposited DLC patterns released from the sacrificial PVA template. The 30-nm-thick DLC was perfectly deposited on the vertical wall of the PVA template. After dissolving PVA, a replicated DLC stamp was obtained, which exhibited an excellent correlation with the Si master stamp. The proposed method was conducted on a line feature with a width of 500 nm.

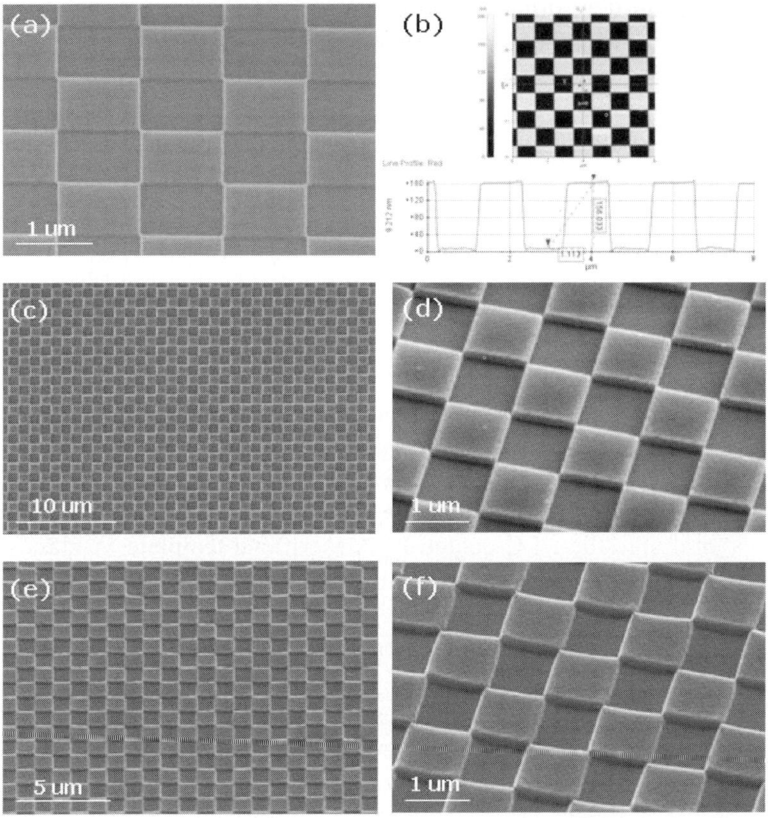

Fig. 8.3 Results of 1-X scale pattern replication using a water-soluble polymer (PVA) and a diamond-like carbon (DLC) coating: (a) SEM image of Si master stamp; (b) AFM image of Si master stamp; (c, d) SEM images of PVA sacrificial template; (e, f) SEM images of replicated DLC stamp. (From Microelectronic Eng., 84, 899, 2007 [11].)

Figures 8.4(a) and 8.4(b) present SEM images of the Si master stamp. It was perfectly replicated using PVA. Figures 8.4(e) and 8.4(f) show that the DLC stamp is in excellent agreement with the Si master stamp. The proposed method was also performed on a line feature with a width of 50 nm.

Figures 8.5 (a-b), (c-d) and (e-f) show the Si master stamp, replicated PVA, and the DLC stamp respectively.

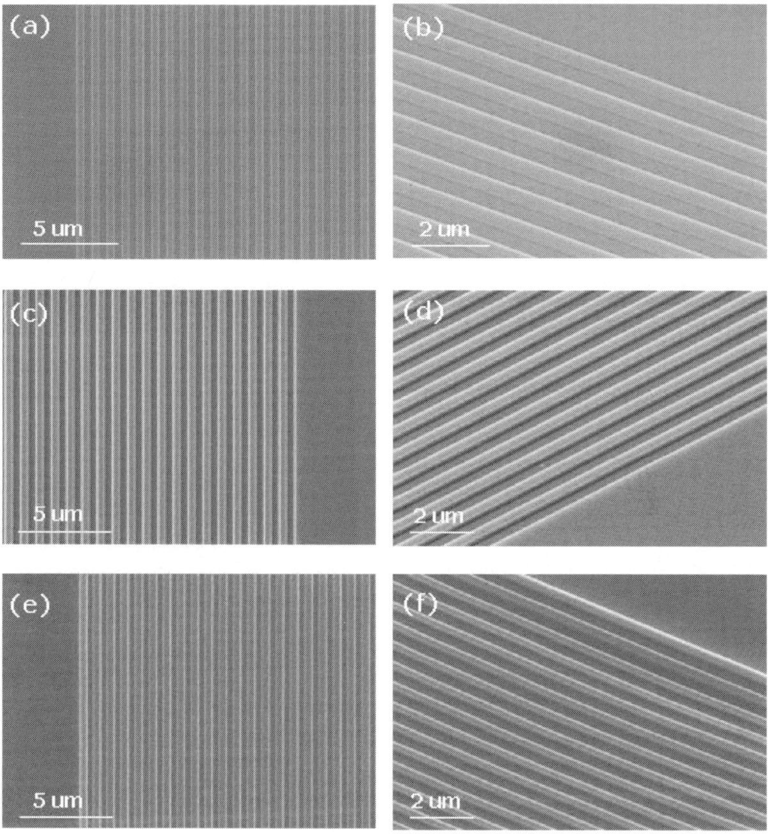

Fig. 8.4 Results of 500-nm scale pattern replication using a water-soluble polymer (PVA) and a diamond-like carbon (DLC) coating: (a, b) SEM image of Si master stamp; (c, d) SEM images of PVA sacrificial template; (e, f) SEM images of replicated DLC stamp. (From Microelectronic Eng., 84, 899, 2007 Ref: 11.)

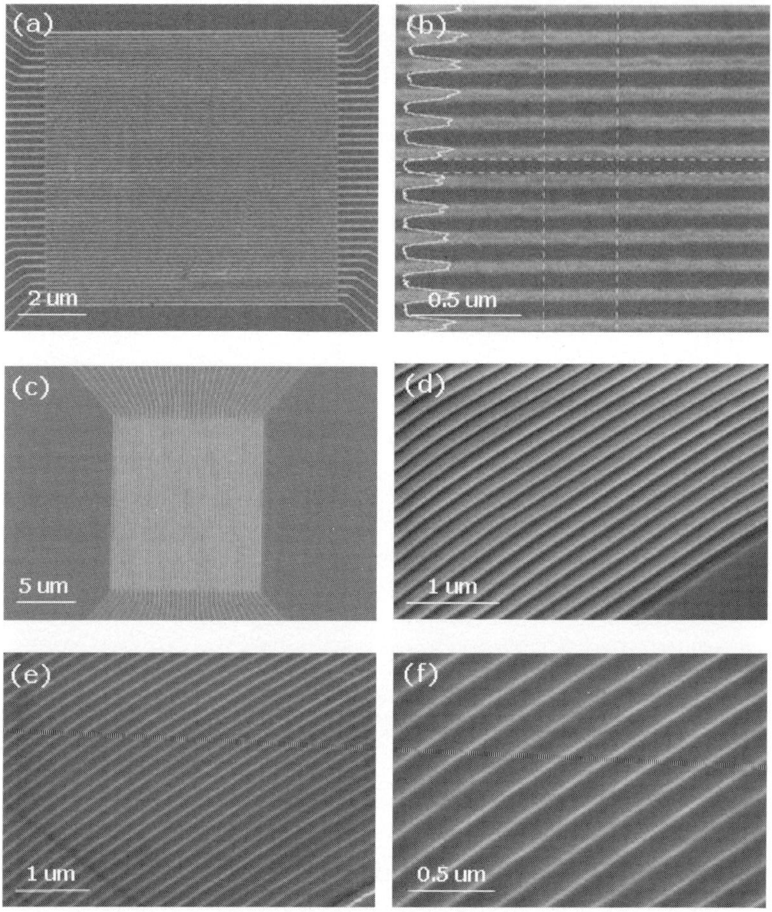

Fig. 8.5 Results of 50-nm scale pattern replication using a water-soluble polymer (PVA) and a diamond-like carbon (DLC) coating: (a, b) SEM image of Si master stamp; (c, d) SEM images of PVA sacrificial template; (e, f) SEM images of replicated DLC stamp.

8.2.3 Stamp fabrication technique using two-photon polymerization and DLC coating

UV-NIL demands only a low pressing pressure and can be done at room temperature, which has some advantages compared to the thermal-type NIL. Until now, most of the research works concerned with NIL are focused on two-dimensional (2D) patterning for mass production. However, the fabrication of three-dimensional (3D) nano/microstructures is of great importance in diverse fields of modern technology for the development of highly functional applications. From this point of view, some 3D nano/microfabrication techniques have been developed using nonphotolithographic approaches. Among them, two-photon polymerization (TPP) is considered a promising technique and has been extensively studied over the last decade in order to fabricate complex 3D micro/nanodevices by a method of stacking layers[12–20]. TPP is considered currently as a unique technique that could create a real 3D microstructure with a high spatial resolution beneath the limit of diffraction of light. More detailed explanations on TPP are given in numerous reports[14–20]. However, long processing time is normally required to create 3D microstructures due to the intrinsic TPP characteristic of the layer-by-layer accumulation. For this reason, TPP is evaluated as an incongruent process for mass production in spite of its unique capability of complete 3D micro- and nanoscale fabrication. Therefore, the increase of throughput in TPP becomes an important issue for its practical use as a nanofabrication process[21–23]. In this work, the fabrication process of a 3D multilayered stamp, applicable to UV-NIL and with a resolution of sub-100-nm scale, is introduced for the fabrication of 3D or multilevel nano/microstructures. The single-step fabrication of multilevel structures has some meaningful advantages of avoiding an alignment issue of several 2D stamps (or masks) including low fabrication cost due to a single stamp and a simple process. Figure 1.6 summarizes the fabrication procedure for 3D multilayered stamp for UV-NIL, which consists of four different parts: a glass plate, polymer patterns, a DLC layer, and a thick glass plate for reinforcement. On a thin glass plate, 3D polymer patterns were created using TPP. For this work, an optical femtosecond laser system has been utilized. The full specifications of the developed system were given in our previous reports[14–16,18]. Through these procedures, 3D polymer patterns on the thin glass plate were created for use as a UV-NIL stamp. This had the advantages of a flexible design, as well as being perfectly transparent to UV light.

Another important issue concerned with TPP that comes into light in recent research work is how to improve the resolution of the microstructures. In general, it is well known that the resolution in TPP is dependent on the process parameters such as laser power and exposing time. The previous works have shown that the high spatial resolution of a voxel could be achieved with a delicate control of the laser dose into the near-threshold energy region for polymerization[13–16]. Moreover, an approach using a radical quencher to improve the resolution of a voxel has been reported in a recent study[25], in which a lateral spatial resolution of 100 nm was achieved using the quenching effect. However, in case the quenching effect is utilized, the mechanical strength of the polymerized structure is inevitably reduced, despite achieving a high spatial resolution, because of the smaller molecular weight of polymerized structures[26]. Therefore, a challenging issue related with the enhancement of the mechanical strength still remains when it comes to the practical use of radical quenchers in 3D nanofabrication.

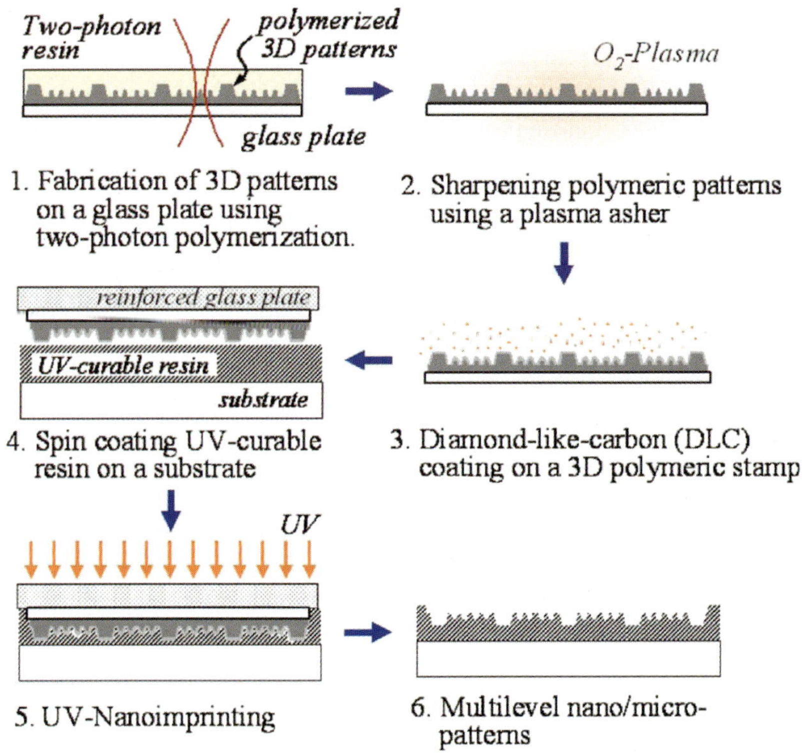

Fig. 8.6 Schematic illustration of the fabrication procedure of the multilayered stamp and the UV-nanoimprint process for the creation of nano/micropatterns in a single step. (From Appl. Phys. Lett., 88, 203105, 2006 Ref: 24.)

When DLC is coated onto polymer structures, the periodic redeposition of anti-adhesion layer is not needed. In addition, the mechanical strength of the polymer patterns is to some extent increased due to the high strength of DLC. In order to evaluate the increase of hardness by DLC coating quantitatively, nanoindentation tests were conducted. Two polymer squares, 1.5 mm in thickness and 3.0 mm in width, were prepared to compare the hardness between one with the DLC coating of 20 nm thickness and the other sample (original polymer square). In the test results, it is seen by observation of the shallow indentation depth (~ 20 nm) that a considerable, fivefold (approximately) increase in hardness was obtained in case the DLC coating was present.

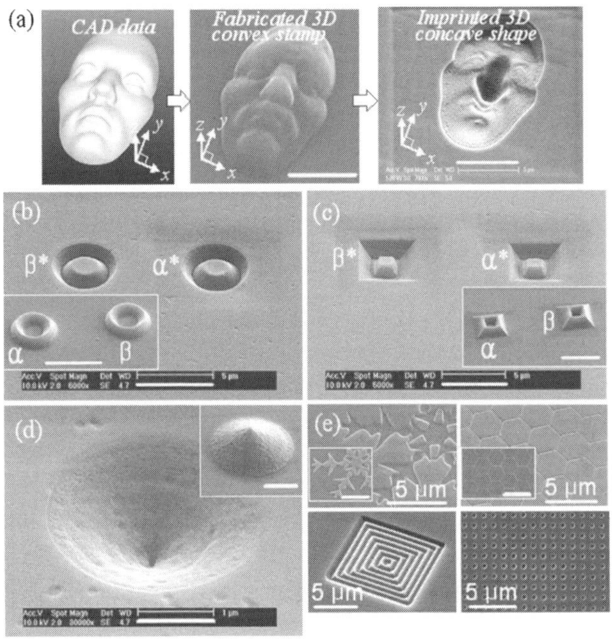

Fig. 8.7 SEM images of some imprinted 3D and multilevel structures; (a) schematic sequential procedure of creating 3D face shape from designed shape to imprinted result: computer-aided design (CAD), SEM image of fabricated 3D convex face shape (*stamp*, the scale bar is 5 μm), and 3D concave shape (*imprinted result*). [(b) and (c)] Imprinted two circular rings and two rectangular rings (α*, β*); the inserts are the image of corresponding multilayered stamps (α, β), and all scale bar is 5 μm. (d) Imprinted cone-shaped microstructure (shown in the insert, the scale bar represents 1 μm). (e) Various imprinted 2D patterns; the inserts are the images of fabricated stamps. (From Appl. Phys. Lett., 88, 203105, 2006 Ref: 24.)

In this work, the multilayered stamps were fabricated with a DLC coating of around 10 nm on superficial polymer patterns. However, more effort is needed to evaluate explicitly the variation of antiadhesive qualities depending on various thicknesses of DLC coating between the stamp and the resist. Figures 8.7(a)–8.7(e) show some fabricated 3D concavoconvex multilayered stamps as well as their corresponding UV imprinted results. The photoresist used in UV-NIL was a homemade UV-curable resin whose major component is tripropylene glycol diacylate (TPGDA) with the imprint conditions of an exposure intensity of 14.4 mW/cm^2, an exposure time of 60 s, and an imprint pressure of 930 mbars. Under these imprinting conditions, the stamp is undamaged, because the mechanical properties of Young's modulus, of 3.5 GPa and a yield stress, of 20–40 MPa are realistic values for many glassy polymers and photoresists[27]. From the results, excellent correlations between 3D mutilayered stamps and imprinted features can be observed without any problem. These results indicate that multilevel nano- and micro- scale structures can be created in a single step using multilayered stamps.

8.3 Fabrication of Diamond-like Carbon Stamps for Ultraviolet Nanoimprint Lithography

Diamond-like carbon (DLC) is a promising stamp material for UV-NIL. It has superior mechanical properties and a wide range of DLC deposition methods exist. DLC has a relatively low surface energy compared to glass and quartz, and DLC-coated stamps have been applied to UV-NIL without requiring an anti-adhesion layer coating. However, because the contact angle of DLC is between 60° and 70°, an anti-adhesion layer may still be required for fine imprinting. In addition, UV-transmittance of DLC is quite low. A 200-nm-thick DLC film has a UV-transmittance of less than 1%. As a matter of fact, in the study mentioned above, the thickness of the coated DLC was less than 10 nm.

Here we present a fabrication method for a fluorine-doped DLC (F-DLC) stamp that can be used for UV-NIL. We fabricated two F-DLC stamps with a high contact angle, high UV-transmittance, and sufficient hardness. The stamps were applied to UV-NIL without an anti-adhesion layer coating.

8.3.1 Deposition of F-DLC thin film

Before fabricating the stamps, we performed preliminary tests to determine the optimum fluorine concentration of F-DLC. Samples with different fluorine concentrations were prepared and their contact angle, UV-transmittance, hardness, and etch rate were measured.

Table 8.1 Deposition power, bias voltage values, and XPS analysis results of the samples. (From Nanotechnology, 17, 4659, 2006. Ref: 28)

Sample	Deposition materials	Power (W)		Bias	XPS data (at%)		
		DC	RF	Voltage (V)	F_2	O_2	N
1	C+PTFE	500	25		9.5	0.9	1.7
2	C+PTFE	500	50	-	24.5	1.8	1.3
3	C+PTFE	500	75	-	27.7	2.5	1.2
4	C+PTFE	500	100	-	28.4	1.0	1.4
5	C+PTFE	500	125	-	26.5	1.6	2.6
6	C+PTFE	500	150	-	29.3	0.6	1.4
7	C+PTFE	500	125	50	33.8	2.2	0.7
8	C+PTFE	500	100	50	37.8	0.7	1.1

After depositing a 10-nm-thick DLC film on each sample, the F-DLC film was synthesized using DC and RF magnetron sputtering. In the process, the vacuum chamber was evacuated to less than 1×10^{-6} Torr and then backfilled with argon to 2 mTorr. A 3-inch-diameter graphite target and PTFE target were mounted on the DC gun and RF gun, respectively. Table 1.1 shows the applied power and bias voltage used to deposit the F-DLC as well as the results of an X-ray photoelectron spectroscopy (XPS) analysis that gave the fluorine, oxygen, and nitrogen content of each sample. The XPS analysis was performed with a VG-Scientific ESCALAB 250 spectrometer and monochromatized Al Kα X-ray source. The charge compensation was carried out with two flood guns while the vacuum level of the main chamber was maintained at 1×10^{-8} Torr. A F-DLC film thickness of 200 nm was obtained for all the samples. The application of a bias voltage was found to increase the fluorine concentration. This phenomenon can be explained with the help of two recent publications. Zhang et al.[29] showed that the sp^3/sp^2 ratio of the DLC increases with the bias voltage when the DLC is synthesized using DC magnetron sputtering. Jiang et al.[30] reported that the increase in the sp^3/sp^2 ratio helps to terminate more C=C bonds (or aromatic rings) with fluorine atoms and to form more chain-like structures, thereby increasing the fluorine concentration of the F-DLC. In that study, the F-DLC films were deposited by reactive magnetron sputtering.

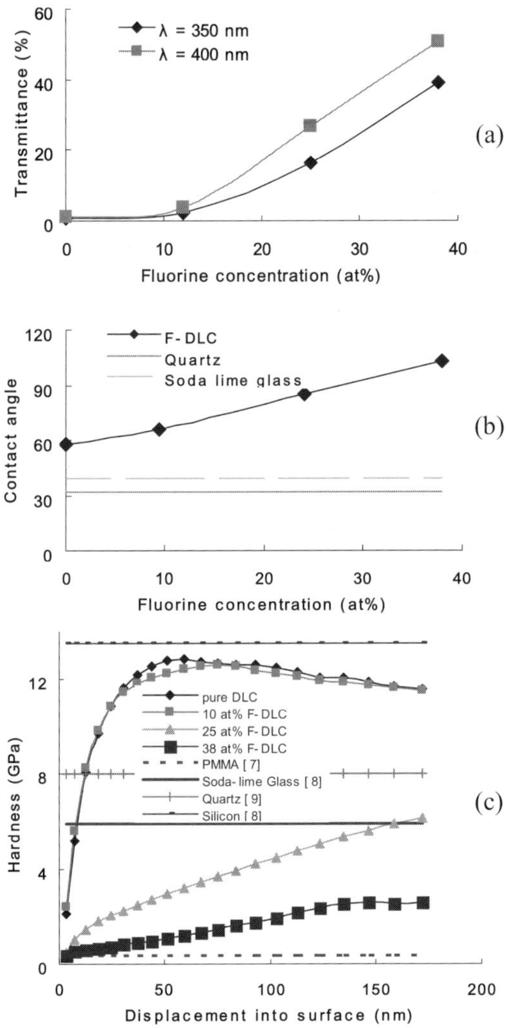

Fig. 8.8 Variation of (a) UV-transmittance, (b) contact angle, and (c) sample hardness with different fluorine concentrations. (From Nanotechnology, 17, 4659, 2006 Ref: 28.)

Tests were performed to obtain the contact angle, UV-transmittance, hardness, and etch rate values of the pure DLC and F-DLC samples (#1, 2, and 8) and samples with a fluorine content of 10, 25, and 38 at%. Figure 8.8 shows the UV-transmittance, water contact angle, and hardness of the test samples as well as supplementary materials[31-33]. The UV-transmittance test was performed with an OPTIZEN 2120 UV Plus UV spectrometer. The hardness was measured using a nanoindenter with a Berkovich diamond indenter. The thickness of the films was 200 nm. The indentation loads were 2.9, 4.8, 6.5, and 6.6 mN, producing indentation peaks at 202, 200, 186, and 182 nm for the 38, 25, and 10 at% F-DLC samples and the pure DLC sample, respectively. These gave corresponding hardness values of 1.9, 4.5, 12.4, and 12.7 GPa, respectively, using the results at an indentation depth of 100 nm.

Etch rate tests were conducted on the samples and on PMMA. The etching was performed at 400 mTorr with flow rates of O_2 and Ar set at 200 sccm and 10 sccm, respectively. The RF power was 200 W during the etching. The diameter of the RIE electrode was 203.2 mm and the distance from the sample to the electrode was 46 mm. While etching, a cover was placed on each sample to obtain unetched surfaces. The etched depth was measured with alpha-step. Many tests were performed using different time periods to obtain the etch rate for each sample. The etch rates of PMMA, pure DLC, and F-DLC samples with 10, 25, and 38 at% fluorine were 130 (±10), 8±1, 10±1, 70±7, and 71±7 nm/min, respectively. The etch rate increased with the fluorine concentration until it reached a saturation point of around 38 at% fluorine.

8.3.2 Fabrication of F-DLC stamp

Our proposed fabrication method is depicted in Figure 8.9. First, a thin layer of DLC without fluorine is deposited on the quartz substrate. This layer is required to provide better adhesion to the substrate. Then an F-DLC layer is deposited. Polymethyl methacrylate (PMMA) is spin-coated on top of the F-DLC layer and used as the etch mask. Sub-100-nm patterns can be etched onto PMMA with e-beam lithography, which is the reason for selecting PMMA as the etch mask. Finally, O_2 plasma etching is used to transfer the patterns to the F-DLC. This method is a direct etching method. When fabricating quartz or glass stamps, the PMMA pattern serves as the mask for the pattern transfer into the chrome. The chrome is then used as the hard mask to transfer the pattern

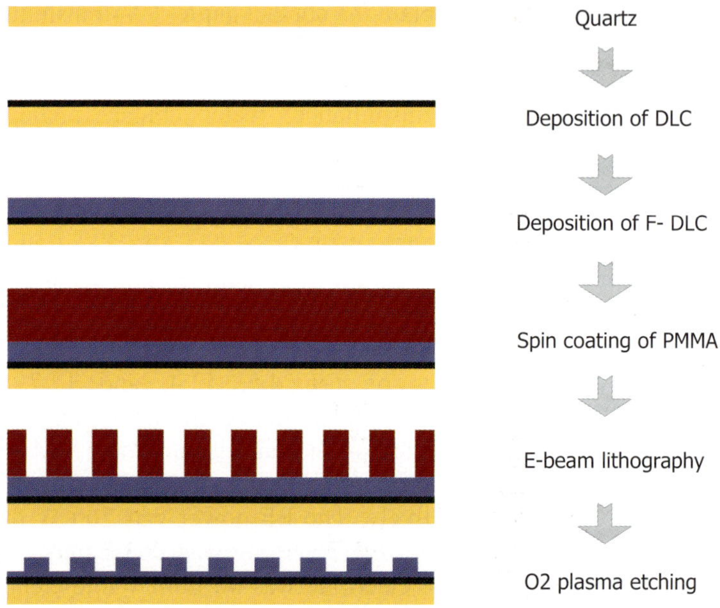

Fig. 8.9 Schematic illustration of the proposed fabrication procedure for a F-DLC stamp.

into the quartz or glass substrate. But in our fabrication method, the patterns on PMMA are transferred directly to the F-DLC so that the etch rate should be comparable to that of PMMA.

Based on the test results, the hardness of the 38 at% F-DLC sample and the contact angle and etch rate of the 10 at% F-DLC sample were too low to be used as a stamp material. The 25 at% F-DLC sample had a high contact angle, UV-transmittance, and etch rate, and sufficient hardness, and therefore was selected as the stamp material. A 200-nm-thick 25 at% F-DLC film was deposited on quartz substrates on top of a thin DLC layer. Then the film was spin-coated with a PMMA layer. Two stamps were fabricated: one with 100-nm line patterns, which had a PMMA thickness of 130 nm, and one with 40-nm line patterns, which had a PMMA thickness of 60 nm. After spin coating, the line patterns were formed on PMMA using e-beam lithography. Finally, the patterns on the PMMA were transferred to F-DLC using O_2 plasma etching. For the stamp with 100-nm line patterns, the etching was performed for 60 s using the same etching parameters as given above. For the stamp with 40-nm line patterns, the etching was performed at an RF power of 25 W and a pressure of 400 mTorr. The parameters were modified for the second stamp to slow down the etching process.

8.3.3 UV-NIL experiment using F-DLC stamp

The polymer pattern was imprinted using an exposure intensity of 30 mW/cm^2, an exposure time of 150 s, an exposure wavelength of 350 to 450 nm, an imprint pressure of 1 bar, and a compression time of 40 s. Figure 8.10 shows results of the atomic force microscope (AMF) measurements of the 100-nm line patterns of the fabricated F-DLC stamp and the imprinted polymer. The height of the stamp and the depth of the imprinted polymer were found to be 74 nm and 67 nm, respectively. The shrinkage of the resin on polymerization was about 12%[4]. Considering the shrinkage ratio of the polymer, it can be stated that the imprinting was performed successfully using the F-DLC stamp.

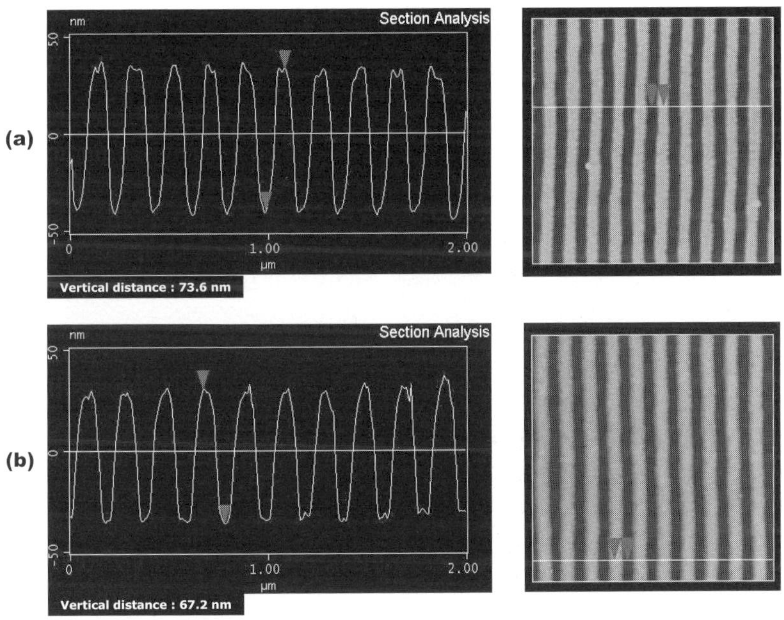

Fig. 8.10 AFM results of the (a) F-DLC stamp and (b) imprinted polymer with 100-nm-wide, 100-nm-spaced, and 70-nm-deep line patterns. (From Nanotechnology, 17, 4659, 2006 Ref: 28.)

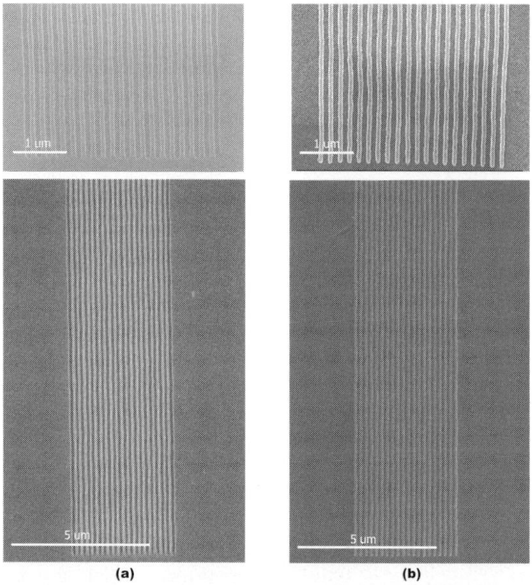

Fig. 8.11 SEM images of the (a) F-DLC stamp and (b) imprinted polymer with 100-nm-wide, 100-nm-spaced, and 70-nm-deep line patterns. (From Nanotechnology, 17, 4659, 2006 Ref: 28.)

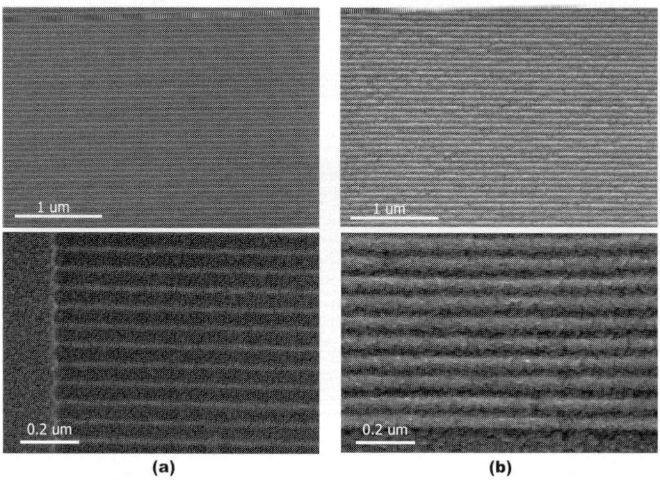

Fig. 8.12 SEM images of the (a) F-DLC stamp and (b) the imprinted polymer with 40-nm-wide, 20-nm-spaced, and 20-nm-deep line patterns. (From Nanotechnology, 17, 4659, 2006 Ref: 28.)

Figure 8.11 shows SEM images of the 100-nm line patterns on the fabricated F-DLC stamp and the imprinted polymer. Figure 8.12 shows 40-nm line patterns on the fabricated stamp and the imprinted polymer. SEM images show that the F-DLC stamps were successfully fabricated and applied to UV-NIL without an anti-adhesion layer coating.

8.4 Summary

We fabricated 100 nm-scale DLC stamps for UV-NIL by using DLC coating and FIB lithography. In addition, stamp fabrication technique using a diamond-like carbon (DLC) coating on polyvinyl alcohol (PVA) was presented. We successfully used the DLC stamps to print wafers with UV-NIL. The dimensions of the features on the stamp correlated well with the corresponding imprinted features. It was demonstrated that a DLC stamp without an anti-adhesion layer can be used for imprinting wafers with UV-NIL.

The applicability of 3D multilayered stamps to a 3D UV-NIL process was demonstrated for the fabrication of 3D or multilevel patterns in a single step. This process has the benefits of cost-effectiveness and mass-production capabilities, as well as overall simplicity. The thin layer characteristic of DLC has some excellent features applicable to 3D UV-NIL. Several 3D nano/micro patterns were created in a single step by the 3D UV-NIL process, and these results demonstrate the possibility of the application of TPP as a multilayered stamping technique.

In the second part, two fluorine-doped diamond-like carbon (F-DLC) stamps with sub-100-nm line patterns were fabricated using a direct etching method. These were applied to ultraviolet nanoimprint lithography successfully without requiring an anti-adhesion layer coating. Tests were performed to determine the optimum fluorine concentration for the F-DLC stamps. The ideal stamp material consisted of 25 at% F DLC with a contact angle of 85°, UV-transmittance of 16.4 to 26.8%, and hardness of 4.5 GPa. The O_2 plasma etch rate of the DLC was increased considerably by the fluorine doping, making it comparable to the etch rate of polymethyl methacrylate (PMMA). Thus, only PMMA was used as the etch mask in the fabrication of the stamps. A larger contact angle provides easier and finer replications of patterns. A higher UV-transmittance makes it possible to reduce either the intensity of the UV light, which increases the life of the UV exposure tool, or the exposure time, which decreases the fabrication time. In addition, a higher

UV-transmittance allows us to fabricate thicker stamps. It would be possible to increase the depth of the patterns if a thicker PMMA layer were applied or the etching process parameters were optimized.

References

1. S. Y. Chou, P. R. Krauss and P. J. Renstrom, *Nanoimprint lithography*, Appl. Phys. Lett. 67, 3114 (1995)
2. M. D. Austin, W. Zhang, H. Ge, D. Wasserman, S. A. Lyon and S. Y. Chou, *6 nm half-pitch lines and 0.04 μm^2 static random access memory patterns by nanoimprint lithography*, Nanotech. 16, 1058 (2005)
3. M. Colburn, S. Johnson, M. Stewart, S. Damle, T. Bailey, B. Choi, M. Wedlake, T. Michaelson, S. V. Sreenivasan, J. Ekerdt, and C. G. Wilson, *Step and Flash Imprint Lithography: A new approach to high resolution patterning*, Proc. SPIE 3676, 379 (1999).
4. E. S. Lee, J. H. Jeong, K. D. Kim, Y. K. Sim, D. G. Choi, J. H. Choi, S. H. Park, T. W. Lim, D. Y. Yang, N. G. Cha, J. G. Park, and W. R. Lee, *Fabrication of Nano- and Micro-Scale UV Imprint Stamp Using Diamond-Like Carbon Coating Technology*, J. of Nanoscience and Nanotechnology 6, 3619 (2006).
5. J. H. Jeong, Y. S. Sim, H. Sohn and E. S. Lee, *UV-nanoimprint lithography using an elementwise patterned stamp*, Microelectron Eng. 75, 165 (2004)
6. C. D. Schaper, *Patterned transfer of metallic thin film nanostructures by water-soluble polymer templates*, Nano Lett. 3, 1305 (2003).
7. T. Ohtake, K. Nakamatsu, S. Matsui, H. Tabata, and T. Kawai, *DNA nanopatterning with self-organization by using nanoimprint*, J. Vac. Sci. Technol. B 22, 3275 (2004).
8. C. D. Schaper and A. Miahnahri, *Polyvinyl alcohol templates for low cost, high resolution, complex printing*, J. Vac. Sci. Technol. B 22, 3323 (2004).
9. F. Pérennès, B. Marmiroli, M. Matteucci, M. Tormen, L. Vaccari, and E. D. Fabrizio, *Sharp beveled tip hollow microneedle arrays fabricated by LIGA and 3D soft lithography with polyvinyl alcohol*, J. Micromech. Microeng. 14, 473 (2006).
10. S. Wang, C. Zeng, S. Lai, Y. Juang, Y. Yang, and L. Lee, *Polymeric Nanonozzle Array Fabricated by Sacrificial Template Imprinting*, J. Adv. Mater. 17, 1182 (2005).

11. K. D. Kim, J. H. Jeong, A. O. Altun, D. I. Lee, D. G. Choi, and E. S. Lee, *Replication of an UV-NIL stamp using DLC coating*, Microelectronic Eng. 84, 899 (2007).
12. H.-B. Sun, V. Mizeikis, Y. Xu, S. Juodkazis, J.-Y. Ye, S. Matsuo, and H. Misawa, *Elastic force analysis of functional polymer submicron oscillators*, Appl. Phys. Lett. 79, 3173 (2001).
13. H.-B. Sun, K. Tanaka, M.-S. Kim, K.-S. Lee, and S. Kawata, *Scaling laws of voxels in two-photon photopolymerization nanofabrication*, Appl. Phys. Lett. 83, 1104 (2003).
14. S. H. Park, T. W. Lim, D.-Y. Yang, H. J. Kong, K. S. Kim, and K.-S. Lee, *Fabrication of nano-precision PDMS replica using two-photon photopolymerization and vacuum pressure difference technique*, Bull. Korean Chem. Soc. 25, 1119 (2004).
15. S. H. Park, T. W. Lim, S. H. Lee, D.-Y. Yang, H. J. Kong, and K.-S. Lee, *Fabrication of microstructures using double contour scanning (DCS) method by two-photon polymerization*, Polymer (Korea) 29, 146 (2005).
16. S. H. Park, S. H. Lee, D.-Y. Yang, H. J. Kong, and K.-S. Lee, *Subregional slicing method to increase three-dimensional nanofabrication efficiency in two-photon polymerization*, Appl. Phys. Lett. 87, 154108 (2005).
17. J. Serbin, A. Egbert, A. Ostendorf, B. N. Chichkov, R. Houbertz, G. Domann, J. Schulz, C. Cronauer, L. Frohlich, and M. Popall, *Femtosecond laser-induced two-photon polymerization of inorganic organic hybrid materials for applications in photonics*, Opt. Lett. 28, 301 (2003).
18. T. W. Lim, S. H. Park, and D. Y. Yang, *Contour offset algorithm for precise patterning in two-photon polymerization*, Microelectron. Eng. 77, 382 (2005).
19. S. Kawata, H.-B. Sun, T. Tanaka, and T. Kenji, *Finer features for functional microdevices*, Nature London 412, 697 (2001).
20. C. A. Coenjarts and C. K. Ober, *Three Dimensional 2-Photon Microfabrication of Silicone Elastomers*, Chem. Mater. 16, 5556 (2004).
21. S. Matsuo, S. Juodkazis, and H. Misawa, *Femtosecond laser microfabrication of periodic structures using a microlens array*, Appl. Phys. A: Mater. Sci. Process. 80, 683 (2005).
22. J. Kato, N. Takeyasu, Y. Adachi, H.-B. Sun, and S. Kawata, *Multiple-spot parallel processing for laser micronanofabrication*, Appl. Phys. Lett. 86, 044102 (2005).

23. C. N. LaFaratta, T. Baldacchini, R. A. Farrer, J. T. Fourkas, M. C. Teich, B. E. A. Saleh, and M. J. Naughton, *Replication of Two-Photon-Polymerized Structures with Extremely High Aspect Ratios and Large Overhangs*, J. Phys. Chem. B 108, 11256 (2004).
24. S. H. Park, T. W. Lim, D. Y. Yang, J. H. Jeong, K. D. Kim, K. S. Lee, and H. J. Kong, *Effective fabrication of three-dimensional nano/microstructures in a single step using multilayered stamp*, Appl. Phys. Lett. 88, 203105 (2006).
25. K. Takada, H.-B. Sun, and S. Kawata, *Improved spatial resolution and surface roughness in photopolymerization-based laser nanowriting*, Appl. Phys. Lett. 86, 071122 (2005).
26. D.I. Bower, *An Introduction to Polymer Physics* (Cambridge University Press, Cambridge, 2002).
27. K. Yoshimoto, M. P. Stoykovich, H. B. Cao, J. J. de Pablo, P. F. Nealey, and W. J. Drugan, *A two-dimensional model of the deformation of photoresist structures using elastoplastic polymer properties*, J. Appl. Phys. 96, 1857 (2004).
28. A. O. Altun, J. H. Jeong, J. J. Rha, D. G. Choi, K. D. Kim, and E. S. Lee, *Fabrication of fluorine-doped diamond-like carbon stamps for UV-nanoimprint lithography*, Nanotechnology 17, 4659 (2006).
29. S. Zhang, X. L. Bui, Y. Fu, D. L. Butler, and H. Du, *Bias-graded deposition of diamond-like carbon for tribological applications*, Diamond Rel. Mat. 13, 867 (2004).
30. M. Jiang and Z. Ning, *Structure evolution of fluorinated diamond-like carbon films prepared at varying source gas flow ratios*, J. Non-Cryst. Sol. 351, 2462 (2005).
31. B. J. Briscoe, L. Fiori, and E. Pelillo, *Nano-indentation of polymeric surfaces*, J. Phys. D: Appl. Phys. 31, 2395 (1998).
32. Y. G. Jung, A. Pajares, R. Banerjee, and B. R. Lawn, *Strength of silicon, sapphire and glass in the subthreshold flaw region*, Acta Materialia 52, 3459 (2004).
33. K. H. Lee and O. Takai, *Nanomechanical properties through nanoindentation method of amorphous carbon and carbon nitride films synthesized by shielded arc ion plating*, Surf. Coat. Technol. 200, 2428 (2005).

9

ZnO NANOWIRES AND NANOBELTS: STRUCTURE SWITCH BY INDIUM DOPING

Hong Jin Fan

Wire or belt-like ZnO one-dimensional nanostructures have been fabricated by various techniques in recent literature. We show that, in a steady-state vapor transport growth mode, the resulting ZnO product under similar growth conditions can be switched between [0001]-axial nanowires and $<11\bar{2}0>$-axial nanobelts simply by adding indium to the source. The former appear as ordered vertical arrays of pure ZnO while the latter are large aspect-ratio belts without spatial ordering. The growth mechanism is intensively discussed based on the examination of their early growth stages and analytical thermodynamic modeling. The calculation result agrees with the experiment observation, that in-situ indium doping influences the nucleation, and supports the dominating growth of *a*-axial nanobelts over *c*-axial nanowires.

9.1 Introduction

Semiconductor nanowires have been regarded as promising build-blocks for semiconductor electronic and optoelectronic circuits. Tailoring the crystallographic orientation of semiconductor nanowires (NW) is one of the key aspects of growth control. It has been shown experimentally that the growth direction can be changed by, e.g. applying corresponding lattice-matching substrates for wurtzite NWs[1,2], or changing growth temperature for ZnSe NWs[3], or using small-sized catalyst for Si NWs [4,5]. Several theoretical studies have been carried out to describe the thermodynamics and kinetics of semiconductor NWs and predict their size-dependent behaviour[5-9].

Since the pioneering work on ZnO nanowires by the Yang group[10] and on metal oxide (including ZnO) nanobelts by the Wang group[11], the synthesis of one-dimensional (1D) nanostructures has been further developed using various techniques. Much attention is being paid to NW synthesis and assembly by developing simple fabrication techniques, designing novel precursors and using various templates[12,13]. In the case of ZnO, a nanowire generally refers to a structure elongated in [0001] direction (viz., *c*-axial) with a quasi-hexagonal cross section; in accordance with the growth habit of wurtzite ZnO crystals. Nanobelts mainly refer to a growth elongated in $<11\bar{2}0>$ (*a*-axial) or $<10\bar{1}0>$ (*m*-axial) directions with rectangular cross sections[11,14].

Vapor-transport-deposition (VTD) inside a horizontal tube furnace is a common method for the growth of low-dimensional nanostructures. Growth of needle-shaped ZnO whiskers were demonstrated as early as 1974 using the VTD method[15]. In recent years, a wide range of 1D semiconductor nanostructures including ZnO have been fabricated with this technique[16,17]. Successful growth is known to be sensitively dependent on various parameters like source type, substrate orientation, temperature, partial pressure of reactants, and catalyst size[18-21].

ZnO nanowires/nanoneedles can be obtained from evaporation and oxidation of Zn[22,23], $ZnAs_2$[24], and $ZnCl_2$[25] at low temperatures (<600 °C), and also by carbothermal reduction of ZnO powder at temperatures of 800–950 °C[10,19,20,26]. The latter technique allows better control of Zn vapor concentration which is important when a patterned growth is required. Nanobelts are generally grown by high temperature (>1000 °C) evaporation of ZnO powder[18]. But the use of other sources in various temperature environments for growth of belt-like structures have also been reported[27-30]. In addition, when impurities are mixed into the source material during the VTD growth, more complex ZnO structures can be developed, for instance, helix[31], rings[32], bridges[33], and nail-wire nanojunctions[34], depending significantly on temperature and the amount of impurities used. Doping of ZnO NWs has also been attempted to enhance the electrical conductivity. However, depending on the doping elements and/or concentration, the ZnO nanostructure can be dramatically different. For example, the Indium-doped ZnO NWs by Xu et al.[35] have an unusual zincblende crystal structure with periodic twinning, whereas in many other cases the indium doping results in a wurtzite-type belt structure[36].

In the common growth mode of carbothermal reduction of ZnO powder, a nearly constant vapor partial pressure and a quasi-equilibrium condition can be assumed so that the growth is almost completely determined by thermodynamics. As a result, the nanowire growth direction is mostly along [0001], in accordance with the crystallographic habits of ZnO. Non-*c*-axial ZnO nanobelts are usually observed under non-steady growth conditions with or without addition of impurities to the source[30,36,37]. Under these conditions, the growth kinetics can be dramatically different, e.g. the mobility of atoms is high enough to smooth up the high-energy side faces and diffuse to lower-energy front faces. As the ZnO (0001) faces are the highest-energy low-index planes (as shown by theoretical calculations[38,39]), a fast growth in [0001] is thermodynamically favored over that in $<11\bar{2}0>$ or $<01\bar{1}0>$. Kong and Wang[14] controlled the growth kinetics (by pressure variation and temperature ramp) so that the energy barrier between (0001) plane and $(11\bar{2}0)$ plane of ZnO was somehow overcome to facilitate its *a*-axial growth.

In this chapter, we demonstrate that in the carbothermal reduction VTD experiments, routinely performed for ZnO NW growth, introducing an additional indium oxide source causes a reproducible transition from usual *c*-axial NWs to *a*-axial nanobelts (NB) while keeping all other conditions constant[40]. The growth kinetics that has been proposed for oxide NBs by high-temperature decomposition and crystallization[14,18] may not explain the variation of shape and crystallographic orientation in this relatively low temperature VLS case. Instead, we show that it is correlated to the indium doping. Furthermore, a thermodynamic model is applied to predict the dimension dependence of the structure/orientation selection. The result of the modeling supports the hypothesis that indium doping plays an essential role in the nucleation of nanobelts, and supports its growth.

9.2 Sample Fabrication

Growth of ZnO nanostructures was conducted inside a double tube furnace system through the VTD method (see Fig 9.1). The growth system comprised of a standard quartz tube furnace (Carbolite, inner tube diameter: 50 mm), vacuum adaptors, a gas flow controller, an oil-free pump station, and an electronically-controlled needle valve. In order to eliminate a possible substrate-induced growth orientation change, as has been demonstrated for Si, GaN, and ZnO nanowires, we used (0001)-oriented GaN substrates decorated with gold nanodot arrays for both experiments. The gold nanodots templates were prepared using a modified nanosphere lithography process[41]. In order to obtain a sufficient Zn vapor pressure, the source materials and substrates were put in a small alumina tube (inner diameter: 15 mm). For the growth of epitaxial nanowire arrays, a mixture of ZnO+C powder (1:1 wt %) was used as the source and located at position I in Fig 9.1 (referred to as "I" experiment). For the growth of nanobelts, an additional alumina boat filled with In_2O_3+C powder (1:1 wt %) was put at position II (referred to as "I + II" experiment). The Au coated GaN/Si substrates were located near the open end of the small alumina tube. A constant Ar gas flow (25 sccm) was directed into the small reaction tube by another small alumina tube. No additional O_2 gas was introduced since the residual/leaky air in our setup, with a partial pressure of ~10^{-3} mbar, provides sufficient oxygen from a thermodynamic point of view. Indeed, when we mix additional oxygen into the pure Ar (~ 1%), the ZnO reduction was significantly suppressed and no growth of nanowire was observed. During the whole process from heating, maintaining to cooling, the Ar gas flow rate and total pressure (200 mbar) were kept constant so as to establish a quasi equilibrium environment. The system was kept at the peak temperature for 25 min, and then cooled down naturally to room temperature. Such a setup has also been used to grow ordered arrays of ZnO nanowires[26,41] and submicron-thick pillars on GaN/Si substrates[42].

9.3 Overview of the Result

A brief overview of the result is shown in Fig 9.1(b). When only the ZnO + C source was used at position I, vertical arrays of ZnO nanowires were obtained growing along [0001] direction and with hexagonal cross

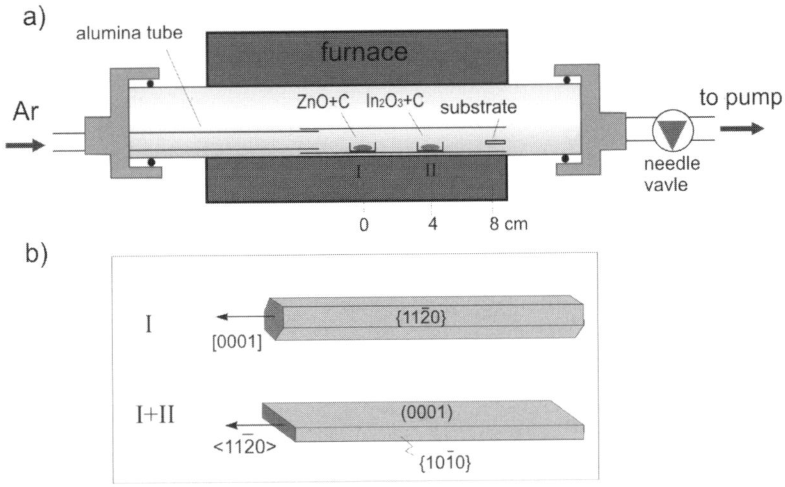

Figure 9.1 (a) Experimental setup for the growth of ZnO 1D nanostructures inside a quartz tube horizontal furnace. Both the source powder and substrate were located inside the reaction alumina tube, where their relative distances are labeled. (b) Schematics of the 1D nanostructure obtained under different growth conditions: *c*-axial nanowires when only ZnO+C source was used at position A; *a*-axial nanobelts when both ZnO+C (at position I) and In_2O_3+C (at position II) sources were used. (Reprinted from Ref. 40.)

sections. However, if both ZnO + C (at position I) and In_2O_3 + C (at position II) were used, or ZnO + C (at position I) only but in an indium-contaminated alumina tube, the major product was randomly oriented high aspect-ratio nanobelts (together with some other irregular belts) which evenly cover the whole substrate surface. Detailed analysis showed that the composition of the nanobelts was ZnO doped with indium (the atomic percentage of In, defined as In/(In+Zn) was 4–6 at%) and the growth was along $<11\bar{2}0>$ directions. A comparison between the nanowires and nanobelts is presented in Table 1. We repeated the experiments more than 10 times and found that it is a reproducible phenomenon. At least two conclusions can be drawn from table 9.1. First, vertically aligned ZnO nanowires within long-range ordering can be easily realized by combining catalyst templates and VTD technique. We established a new method for growing $<11\bar{2}0>$-axial ZnO nanobelts at relatively low source temperatures of 860 °C compared to 1300–1400 °C used by Kong[14]. Second, the sharp difference between the results of

"I" and "I + II" experiments implies that indium plays a major role in defining the crystallographic orientation of the nanobelts, in addition to providing the dopant. In the following we examine the two structures in more detail.

Table 9.1 Comparison of the results from "I" and "I + II" experiments.

	"I" experiments	"I + II" experiments
Sample morphology	Vertical array of nanowires	Randomly oriented nanobelts
Composition	Pure ZnO	ZnO doped with In
Growth direction	[0001]	$<11\bar{2}0>$
Growth rate	Low	High
Length-to-width ratio	10–40	~120

9.4 [0001]-axial Nanowires

Figure 9.2 demonstrates examples of the ZnO nanowire arrays. The vertical alignment and periodic ordering of the wires was found to be present on a large scale for both cases. Mainly, a single wire grows from each Au nanodot site. Some stripes and empty sites are present in analogy to defects in a crystal lattice, which is mainly due to the size inhomogeneity and evaporation of the polystyrene spheres. The hexagonal ends of the wires (see inset) indicate that their main axis is along [0001]. The mean diameter of the ordered nanowires is ~ 50 nm. The nanowires are completely perpendicular to the GaN surface, as expected for epitaxial growth. As proved before[43], the GaN(0001) layers are ideal substrates for the fabrication of highly ordered *c*-axial ZnO nanowires.

In our experiments, the use of the small tube and relatively low substrate temperatures was the key to the successful growth of arrays of single nanowires at individual Au sites. The small alumina tube works as a spatial confinement to maintain a sufficiently high local vapor concentration near the substrates so that the temperature could be effectively lowered to reduce the thermal-induced diffusion of small Au dots (≤30 nm). In this way, the separation of the Au disks into smaller dots can be avoided, and hence, only one wire grows out of the disk.

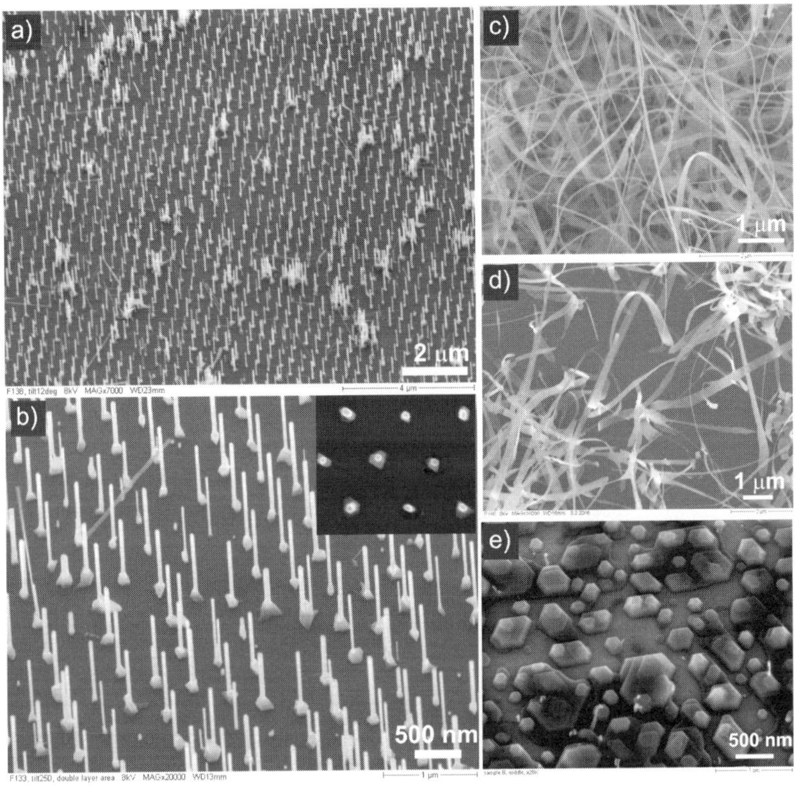

Figure 9.2 SEM images of the structures from experiments where source types were used. (a-b) Pure ZnO+C source located at position I (i.e., experiment I). The inset is the corresponding top view of the wires showing their quasi-hexagonal ends. (c) ZnO + C at position I and In_2O_3 + C at position II (i.e., experiment I + II). (d) ZnO + C source at position I in a indium-contaminated tube. (e) ZnO + In_2O_3 + C (1:1:1 wt%) source located at position I.

The single crystallinity and VLS epitaxial growth of the ZnO nanowires were verified by cross sectional TEM analysis. Figure 9.3(a) shows a low-magnification view of two ~ 30 nm diameter nanowires. In addition to the vertical alignment, one can readily see the Au particles present at the tips of the nanowires. This confirms the growth mechanism to be VLS. The SAED patterned in Fig 9.3(b) and the HRTEM image in Fig 9.3(c) confirm that the nanowires are wurtzite single crystals with the main axes along the [0001] direction. In Fig 9.3(c), the crystal perfection and atomic arrangement of both ZnO and Au, as well as the orientation relationship of Au (111) // ZnO (0002), can be identified. The roughness of the nanowire surface is no more than four atomic layers of the coverage. The lattice distance for the (0002) plane determined here is 0.52 nm, comparable to that of bulk ZnO.

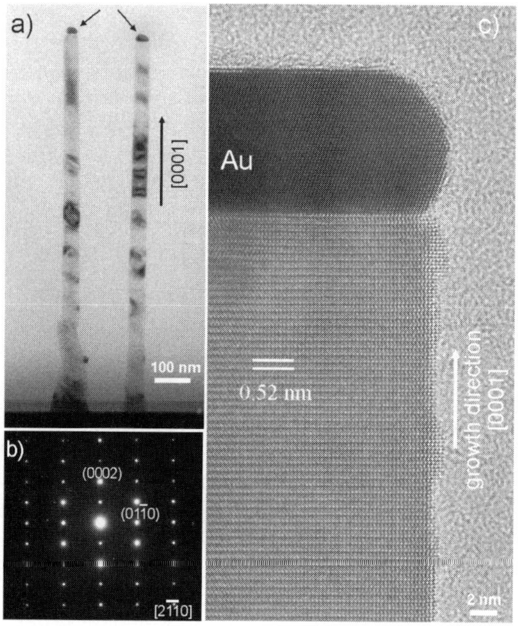

Figure 9.3 TEM images of the nanowires. (a) A cross sectional view of two nanowires, showing their vertical alignment and the presence of Au particles on the wire tips. (b) SEAD pattern recorded from one nanowire, confirming the axes direction along [0001]. (c) HRTEM image recorded near the tip of one wire, revealing the lattice structures of both Au and ZnO. (Reprinted from Ref. 40.)

9.5 $<11\bar{2}0>$ -axial Nanobelts

In "I + II" experiments, the position of the In_2O_3 source was carefully chosen, so that on the one hand a small amount of In vapor could be produced while on the other hand the possible formation of ZnO-In_2O_3 complex superstructures[33] was avoided. Figure 9.2(b) shows example SEM images from the results of a number of repeated "I + II" experiments. These nanobelts have a typical length of ~ 20 μm after a 25 min growth; nearly ten times longer than the nanowires when only a ZnO source was used. These have widths in the range of 100–220 nm. Most of the nanobelts are smooth along their lengths, but a small amount of zigzag nanobelts were also observed. In fact, we realized that the formation of the nanobelts is substrate independent: we found very similar results when using different substrates of Si, SiO_2, and GaN while otherwise under the same growth conditions. Growth of the nanobelts was also catalyzed by gold, since no growth was observed at places without Au covering, and catalyst nanoparticles were indeed observed at the tips of the belts. Unlike the vertical nanowires in Fig 9.2(a), the nanobelts uniformly cover the whole substrate surface with random orientations. This is unsurprising since the growth direction of the nanobelts is along $\left[11\bar{2}0\right]$ so that no epitaxial growth is expected on the (0001)-oriented GaN surfaces. Figure 9.2(e) is the result when a ZnO + In_2O_3 + C (1:1:1 wt %) powder mixture was used. It appears that indium has dissolved into the gold, forming laterally extended quasi-hexagonal pads which is believed to be an In-Au-Zn ternary alloy. No growth of ZnO nanostructure was observed, probably because the growth was suppressed due to the high indium concentration in the alloy.

A TEM image of three smooth nanobelts is shown in Fig 9.4(a), from which the thickness is estimated to be about 10 nm. The SAED pattern (inset in Fig 9.4(a)) confirms that the nanobelts are single crystals growing along $<11\bar{2}0>$, with $\pm(0001)$ top and bottom surface, and $\pm(01\bar{1}0)$ side surfaces. The nanobelts grew also via VLS mechanism, as indicated by the Au tips (inset of Fig 9.4(a)). EDX spectra (Fig 9.4(b)) taken from randomly selected belts show the presence of In, 4–6 at%, besides the major content of Zn and O, as well as Au near the tip shown in the inset of Fig 9.4(a). This implies that In atoms are doped within the ZnO nanobelts. The HRTEM image in Fig 9.4(c) shows clearly the smooth and sharp surface, and its single-crystal hexagonal structure. The nanobelts in the present work are free of twins or dislocations, as consistent with the undoped *a*-axial ZnO nanobelts reported by Kong[14] but different from the nanorings[44] and bicrystal nanobelts[45,46] in which local segregation of impurity ions induces planar defects. Of particular note is the fact that the atomic percentage of indium in Ref. 44 was nearly 50%, much higher than that 4–6 % of the nanobelts reported here. This suggests that in our case the In dopant substitutes Zn lattice sites.

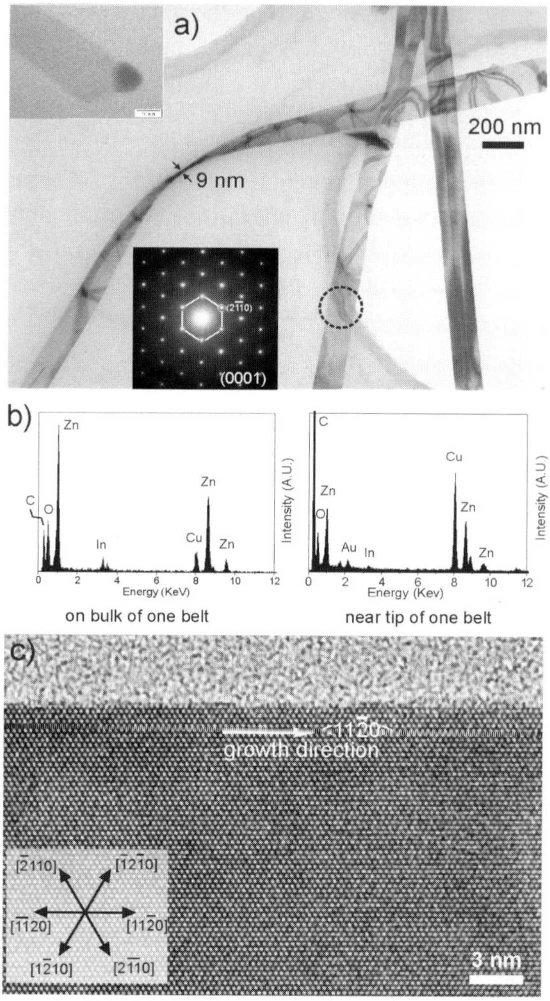

Figure 9.4 TEM images of the straight nanobelts. (a) An overview of three nanobelts. Top inset: one nanobelt showing the Au tip. Bottom inset: SAED pattern recorded from the circled area. (b) EDX spectra. Left: from the circled area in (a); right: near the tip (inset of (a)). The carbon and copper signals are from the TEM grid. (c) HRTEM image of part of one nanobelt, showing its single-crystallinity and atomically clean surface, as well as the growth direction in $<11\bar{2}0>$. The inset illustrates the six equivalent directions within the (0001) plane. (Reprinted from Ref. 40.)

As for the zigzag nanobelts, these are also *a*-axial single crystals. TEM examination (data not shown) shows that the angles at the kinks are always 120°, at which the belts switched to equivalent $<11\bar{2}0>$ crystallographic directions while maintaining the atomic coherency without involving twins or stacking faults. The top surfaces of the belts are (0001) planes. EDX spectra indicate that these zigzag belts are also In doped. Interestingly, these zigzag nanobelts are similar in structure and orientation to the 2D ZnO nanodendrites, which were formed by oxidation of polyhedral Zn microcrystals in air[47,48].

9.6 Early Growth Stages

Based on the above results, it appears that a switching from c-axial wires to a-axial belts is related to In doping. In order to prove this, we examined early growth stages of both structures. Figure 9.5(a1) shows the result of a similar "I" experiment as in Fig 9.2(a-b), except for a shorter growth time (~ 5 min).

Individual pyramids of ZnO are seen, with some short rods tending to grow up vertically out of the substrate surface. This indicates a heteroepitaxial nucleation of ZnO on GaN, so that the pyramids and the subsequent nanowires are [0001] orientated. In contrast, the ZnO nanobelts in their early growth stages show a significantly different morphology. Figure 9.5(b1) is the result after a 15 min "I+II" experiment. Nearly no vertical-aligned 1D structure was observed. Instead, two-dimensional nucleation seems to be favored at the beginning, forming quasi-hexagonal pads with much larger size than the pyramids in Fig 9.5(a1). Sheet-like structures then extend out laterally from the sides of the pads. An example of the embryo of the nanobelts is seen in Fig 9.5(c), which shows clearly the starting point of the belt structure from the side of the hexagonal pads. Interestingly, the whole embryo is a single-crystal entity as seen from the corresponding diffraction pattern in Fig 9.5(c) as well as from HRTEM examinations (data not shown). EDX spectrum also reveals the presence of indium in the embryo with a similar atomic percentage as in the nanobelts.

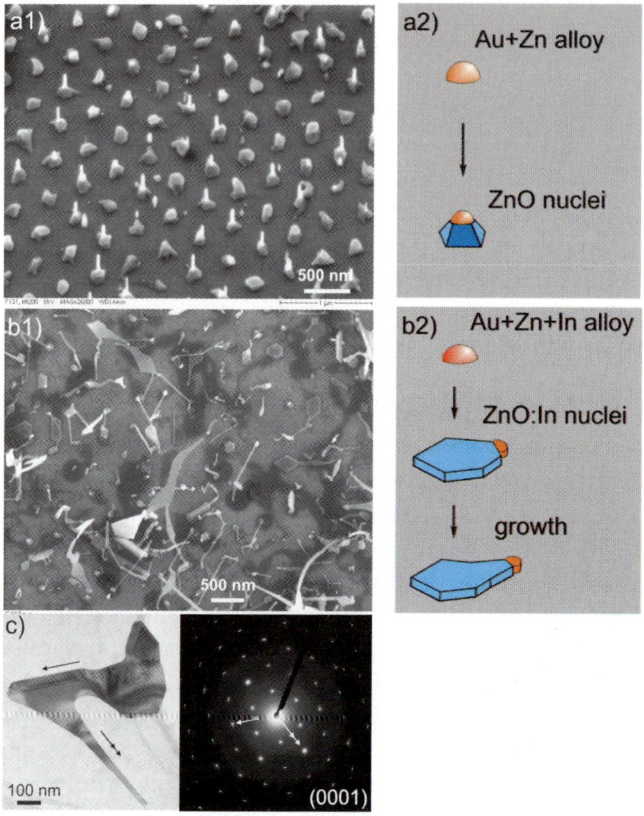

Figure 9.5 The early growth stage of the ZnO 1D nanostructures using separated Au nanoparticles as growth templates and catalyst. (a1) 20° tilted view of the sample after a 5 min "A" experiment. (b1) Top view of the sample after a 15 min "A+B" experiment. (c) TEM image and the corresponding diffraction pattern of an embryo in (b). The arrows indicate the growth orientation of the branches in equivalent $<11\bar{2}0>$ directions. (a2) Schematics of the early growth stage of the c-axial nanowires. (b2) Schematics of the early growth stage of the a-axial nanobelts. The main difference is that the nanowire grows upwards on the top (0001) plane of a ZnO columnar/pyramidal nuclei, whereas the nanobelt grows mainly from the side faces of a quasi-hexagonal ZnO:In pad (dimensions not to scale).

9.7 Discussion on the Growth Mechanism

For the bottom-up vapor-phase growth of 1-D structure of nanomaterials (especially defect-free single crystals), the most widely accepted growth models are VLS and vapor-solid (VS). In VLS, the driving force for the growth is the precipitation and nucleation at the liquid-solid interface, in which the size of the growing 1-D structure depends on the composition and the size of the liquid droplet. In VS, the 1-D growth is mainly controlled by kinetics, for which the temperature and the supersaturation ratio are two dominant processing factors in affecting the morphology of the products. Both mechanisms have been widely suggested for the growth of ZnO nanowires and nanobelts. In our study, since the Au is mandatory for the formation of both nanostructures and an Au tip does appear at their ends, we believe that VLS is the dominant growth mechanism. For the vertical-aligned wire, the growth seed is a (0001)-oriented ZnO pyramid or a hexagonal column which is formed through Zn precipitation and oxidation (see Fig 9.5(a1)). However, the nanobelt starts from the side faces of a hexagonal-shaped seed and proceeds under the guidance of the alloy tip (see Fig 9.5(b1)).

In order for a crystal to switch from one crystallographic structure or orientation to another, a key parameter is the difference of Gibbs free energy[49]. For example, the energy needed to change from zincblende CdTe to wurtzite can be over 10 meV per atom[50]. As for silicon, it has been shown that the sum of surface energy and edge tension at the liquid-solid interface of a <110>-axial nanowire is lower than that of <111>-axial one when the nanowire diameter is smaller than ~ 11 nm[5,8]. As the ZnO(0001) faces are the highest-energy low-index planes[38,39], a fast growth in [0001] is thermodynamically favored over $<11\bar{2}0>$.

Our result reveals clearly that a trace amount of indium in our experiments is still enough to alter the morphology and growth direction of ZnO nanostructure from an energetically favoured [0001] to $[11\bar{2}0]$ direction. The early growth stages (Fig 9.5) show the different nucleation behavior in the presence of In. Indium vapor generated by carbothermal reduction of In_2O_3[51] can dissolve into Au up to a solubility of ~ 24 at% at about 650 °C. Nevertheless, the In content in Au is expected to be much lower than Zn, since the In vapor concentration is much lower than that of Zn (We did not observe any growth in "II" experiments when only In_2O_3+C source was used). Therefore, we propose that a Au-Zn-In ternary phase gives rise to a certain strain to change the surface free energy so that the nucleation behavior of ZnO at the solid-liquid interface upon supersaturation and oxidation of Zn is modified. Such ternary alloy continues at the tip of the nanobelt during its elongation (see Fig 9.5(b1)), so as to maintain a stable belt-structure. As for the kinking of the zigzag belts, it is probably due to some thermal or strain instability at the liquid-solid interface, which can occur without involving much energy since the free energy is the same for the equivalent $\{11\bar{2}0\}$ planes. Similar kinking in equivalent lattice planes was also observed in Si nanowires[5].

Furthermore, as discussed by Jie et al.[29] for their In-doped ZnO nanobelts and Kar et al.[52] for their nano- to micrometer-wide ZnS belts, direct impingement of Zn and In vapor atoms onto the side faces of the growing nanobelts could also occur, giving rise to lateral growth on the $\{10\bar{1}0\}$ faces via the VS mechanism. This explains why some of the nanobelts have widths larger than the catalyst Au tip.

9.8 Thermodynamic Modeling

The lateral extended nuclei due to the incorporation of In into the Au and Zn alloy are certainly not thermodynamically stable (nor kinetically preferable) if composed of pure ZnO. The question is *whether the post-nucleation growth of belt structure is also due to the presence of indium.* In order to answer this, we examine the relative stability of the nanowires and nanobelts with dimensions of interest, in the absence of indium[53].

The analytical, shape-dependent thermodynamic model used here is based on a geometric summation of the Gibbs free energy[56]. This model has proved to be successful in describing the shape of nanocrystals and nanobelts[54,55,56]. As the ZnO nanostructures under consideration here are relatively large, a "truncated" version (edge and corner effects are ignored due to the large feature sizes studied here) of the model is used. The total free energy G is described in terms of the surface energy γ_i for each crystallographic surface i, weighted by the factors f_i (such that $\Sigma fi = 1$) as,

$$G = \Delta_f G^o + \frac{M}{\rho}(1-e)\left[q\sum_i f_i \gamma_i\right] \tag{1}$$

where $\Delta_f G^0$ is the standard free energy of formation of the bulk (macroscopic) material, M is the molar mass, q is the density and e is the volume dilation induced by the surface stresses (as defined in Ref.57). Since there is no experimental data available for the specific surface energies of ZnO, we used the *ab initio* (all-electron) surface energies of $\pm\{10\bar{1}0\} = 2.3\ \mathrm{J{\cdot}m^{-2}}$, $\pm\{11\bar{2}0\} = 4.1\ \mathrm{J{\cdot}m^{-2}}$ and $\pm(0001) = 5.4\ \mathrm{J{\cdot}m^{-2}}$ calculated by Wander and Harrison with the hybrid B3LYP density functional[58]. It is important to point out that these energies were calculated at zero temperature, which has implications for the applicability of the model results to real experiments. Also, our model results are more related to low-temperature quasi-equilibrium conditions and thus have limited applicability to cases where pure ZnO nanobelts are grown via non-steady mass transport.

The morphology of the most common 1-D ZnO structures is geometrically defined in terms of the length and cross-sectional area (see Fig 9.6), assuming a hexagonal and rectangular cross-section for a nanowire and a nanobelt, respectively.

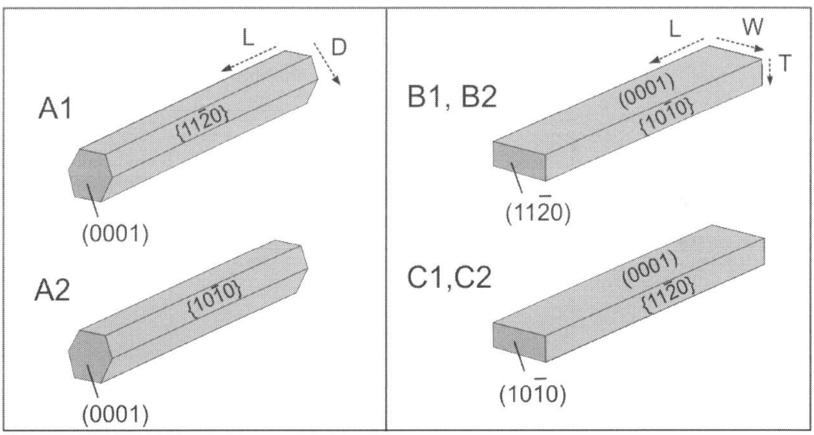

Figure 9.6 Schematic representations of the A1 and A2 nanowires, and B1, B2, C1 and C2 nanobelts, showing orientations and geometric dimensions. "L" refers to length, "W" to width, "T" to thickness, and "D" to diameter. (Reprinted from Ref. 53.)

Type A1 represents the nanowire observed here, type A2 represents an alternative nanowire structure oriented in the ±[0001] direction but enclosed by $\{10\bar{1}0\}$ facets, type B1 and B2 represent the upper and lower limits of the nanobelts reported here and type C1 and C2 represent the upper and lower limits of the nanobelts observed e.g. by Jie et al.[29] In each case, the model was used to calculate the total free energy as a function of total volume (in terms of the number of ZnO formula units) in Fig 9.7(a) and aspect ratio in Fig 9.7(b).

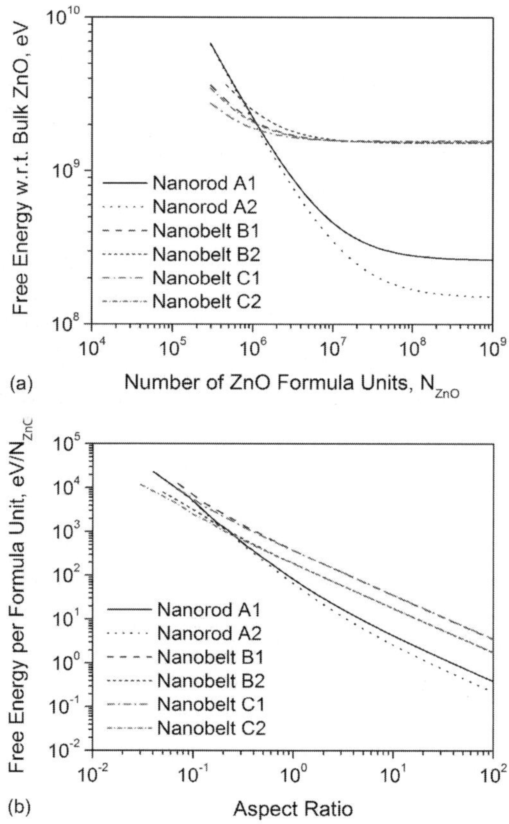

Figure 9.7 Free energy of the nanowire and nanobelt shapes as a function of (a) total volume in terms of the number of ZnO formula units, and (b) aspect ratio. The change of aspect ratios is made by lengthening along the corresponding principle axis. (Reprinted from Ref. 53.)

In Fig 9.7, all changes are in response to extension along the principle axis (i.e. lengthening). We can see from Fig 9.7(a) that at large volumes, pure ±[0001] oriented ZnO nanowires are clearly thermodynamically preferred over $\pm[11\bar{2}0]$ or $\pm[10\bar{1}0]$ oriented nanobelts. There is a cross-over region but turning to Fig 9.7(b) we can see that this cross-over corresponds to low aspect-ratio nanostructures (aspect ratio < 1) such as nanoplatelets. Over an aspect ratio of ~1 (corresponding to a quasi zero dimensional nanocrystal), pure ZnO nanobelts are not energetically preferred at any size or aspect ratio. This implies that the stability of nanobelts over nanowires in our indium-contamination experiment cannot be thermodynamically accounted for in the absence of indium. In addition, Fig 9.7 also shows that the A1 type nanowires reported here are predicted to be less stable than the A2 type with {10-10} majority side surfaces. This is likely due to an effect of the surface-interface energy of the Au + In + Zn ternary liquid alloy catalyst like the case with Si[5] and ZnSe nanowires[4], which is not included in our model.

In the model calculations presented above, the comparisons between nanowires and nanobelts were made by changing their lengths along ±[0001] directions for nanowires and $\pm[11\bar{2}0]$ direction for nanobelts. Hence, they don't account for the different crystallographic orientations inherent during lengthening. In the following, we compare the stability of each nanostructure, of both pure and In-doped ZnO, that is extended in the same growth direction.

The results of such comparison are depicted in Fig 9.8, showing extensions in the (a) [0001], and (b) $[11\bar{2}0]$ directions, respectively. Note that average values of the nanobelt width and thickness are used for defining the aspect ratio so that the nanowire aspect is ~ 30, and the nanobelt aspect is ~ 400. If $[11\bar{2}0]$ growth dominates, the morphology predicted for pure ZnO nanostructures is in agreement with those observed in the indium-doped case. The crossover occurs at a nanobelt length of 4 μm, or a thickness-to-width ratio of ~ 22. Beyond this length, the nanobelts are energetically preferred over nanowires and their growth would be thermodynamically encouraged in the absence of indium.

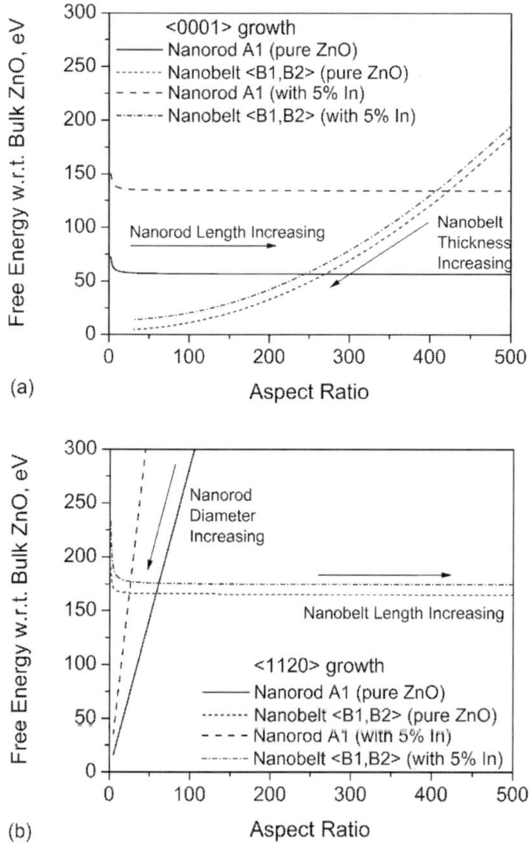

Figure 9.8 Free energy of the nanowire and nanobelt shapes as a function of the aspect ratio for dominated growth (a) along <0001>, and (b) along <11-20>. Average values of the nanobelt width and thickness are used for defining the aspect ratio, so that the nanowire aspect is ~ 30, and the nanobelt aspect ~ 400. The estimated values for In-doped nanostructures are also shown, with 5% In incorporated preferentially into the $\{11\bar{2}0\}$ and $\{10\bar{1}0\}$ non-polar surfaces. (Reprinted from Ref. 53.)

However, if [0001] growth dominates then the morphology predicted for pure ZnO nanostructures disagrees with observations in our indium-doping experiment. High aspect-ratio pure ZnO nanobelts are metastable with respect to a morphological transformation to a nanowire. The cross-over occurs at a nanobelt thickness of 31 nm, much larger than those observed in our experiments (~ 10 nm). Therefore, the results of theoretical modeling indicate that in order for the large aspect-ratio nanobelts to be thermodynamically preferred over the nanowires, indium should promote growth in the $[11\bar{2}0]$ direction and inhibit it in the [0001] direction.

As an estimate of the impact that indium incorporation may have on the relative stability of these nanostructures, the above modeling was repeated with substitutional In defects included artificially in the pure ZnO surface energies. The "defect energy" was approximated by comparing known values for the Zn–O and In–O bond enthalpies[59] and free atom enthalpies[60]. It was also assumed that Zn or In atoms at the surface will form three bonds with surrounding oxygen atoms and will have one dangling bond (taken as half of a Zn–O or In–O bond). Assuming that In atoms incorporate preferentially into the non-polar $\{11\bar{2}0\}$ and $\{10\bar{1}0\}$ surfaces, and that the defect density is approximately 5%, we can estimate the free energy of In-doped nanostructures. Results are also shown in Fig 9.8 for extensions in both [0001] and $[11\bar{2}0]$ directions. We can see that the artificial indium defects have a greater impact on the free energy of the nanowires than of the nanobelts and in Fig 9.8(a) the cross-over occurs at an aspect of ~410 which is consistent with the experimental observations. In general, although this result supports the formation of high aspect nanobelts over nanowires due to indium incorporation, a more definite verification requires explicit and accurate data of surface energies of In-doped ZnO.

9.9 Conclusions

We demonstrated that in the vapor-transport-deposition growth of ZnO nanostructures in-situ indium doping (at 4-6 at %) causes structural change in ZnO from the usual [0001]-axial nanowires to $[11\bar{2}0]$-axial nanobelts. Both are defect-free single crystals grown via VLS mechanism using catalyst Au templates. By studying their early growth stages, we proposed that indium plays a significant role in defining the structure and crystallographic orientation of the nanobelts. This comes about through indium influencing the alloying and nucleation at the liquid-solid interface of the growth front. An analytical thermodynamic model was used to predict the dimension dependence of the shape by comparing the total Gibbs surface energies. The modeling result corroborates our hypothesis that indium doping influences the nucleation and supports the growth of *a*-axial nanobelts over *c*-axial nanowires. Our future work is to thermodynamically model the mechanism through which the atomic arrangement is disordered by indium doping. We would also investigate the quantity of indium needed for a crystallographic transition during nucleation.

Acknowledgments

Contributions and support from B. Fuhrmann, R. Scholz, A. Berger, A. Dadgar, A. Barnard, U. Gösele, and M. Zacharias are greatly acknowledged.

References

1.T. Kuykendall, P. J. Pauzauskie, Y. Zhang, J. Goldberger, D. Sirbuly, J. Den-linger, and P. Yang, Nature Mater. 3, 524 (2004)
2. B. Nikoobakht, C. A. Michaels, S. J. Stranick, and M. D. Vaudin, Appl. Phys. Lett. 85, 3244 (2004)
3. Y. Cai, S. K. Chan, I. K. Sou, Y. F. Chan, D. S. Su, and N. Wang, Small 3, 111 (2007)
4. Y. Wu, Y. Cui, L. Huynh, C. J. Barrelet, D. C. Bell, and C. M. Lieber, Nano lett. 4, 433 (2004)
5. V. Schmidt, S. Senz, and U. Gösele, Nano Lett. 5, 931 (2005)
6. T. Y. Tan, N. Li, and U. Gösele, Appl. Phys. Lett. 83, 1199 (2003)
7. Q. X. Liu, C. X. Wang, N. S. Xu, and G. W. Yang, Phys. Rev. B 72, 085417 (2005)
8. C. X. Wang, M. Hirano, and H. Hosono, Nano Lett. 6, 1552 (2006)
9. Z. Chen and C. Cao, Appl. Phys. Lett. 88, 143118 (2006)
10. M. H. Huang, S. Mao, H. Feick, H. Q. Yan, Y. Wu, H. Kind, E. Weber, R. Russo, and P. Yang, Science 292, 1897 (2001)
11. Z. W. Pan, Z. R. Dai, and Z. L. Wang, Science 291, 1947 (2001)

12. Y. Xia, P. Yang, Y. Sun, Y. Wu, B. Mayers, B. Gates, Y. Yin, F. Kim, H. Yan, Adv. Mater. 15, 353 (2003)
13. H. J. Fan, P. Werner, and M. Zacharias, Small 2, 700 (2006)
14. X. Y. Kong and Z. L. Wang, Nano Lett. 3, 1625 (2003)
15. T. Matsushita, K. Dodaira, J. Saito, and R. Yoshida, J. Cryst. Growth 26, 147 (1974)
16. G. C. Yi, C. Wang, and W. I. Park, Semicond. Sci. Technol. 20, S22 (2005)
17. Z. Y. Fan and G. C. Lu, J. Nanosci. Nanotech. 5, 1561 (2005)
18. Z. R. Dai, Z. W. Pan, and Z. L. Wang, Adv. Funct. Mater. 13, 9 (2003)
19. P. D. Yang, H. Yan, S. Mao, R. Russo, J. Johnson, R. Saykally, N. Morris, J. Pham, R. He, and H. J. Choi, Adv. Funct. Mater. 12, 323 (2002)
20. D. Banerjee, J. W. Lao, D. Z. Wang, J. Y. Huang, D. Steeves, B. Kimball, and Z. F. Ren, Nanotechnology 15, 404 (2004)
21. J. Song, X. Wang, E. Riedo, and Z. L. Wang, J. Phys. Chem. B 109, 9869 (2005)
22. H. J. Fan, F. Bertram, A. Dadgar, J. Chirsten, A. Krost, and M. Zacharias, Nanotechnology 15, 1401 (2004)
23. Y. K. Tseng, C. J. Huang, H. M. Cheng, I. N. Lin, K. S. Liu, I. C. Chen, Adv. Funct. Mater. 13, 811 (2003)
24. H. Chik, J. Liang, S. G. Cloutier, N. Koulin, and J. M. Xu, Appl. Phys. Lett. 84, 3376 (2004)
25. C. X. Xu, X. W. Sun, Z. L. Dong, M. B. Yu, T. D. My, X. H. Zhang, S. J. Chua, and T. J. White, Nanotechnology 15, 839 (2004)
26. H. J. Fan, W. Lee, R. Scholz, A. Dadgar, A. Krost, K. Nielsch, and M. Zacharias, Nanotechnology 16, 913 (2005)
27. Y. B. Li, Y. Bando, T. Sato, and K. Kurashima, Appl. Phys. Lett. 81, 144 (2002)
28. Y. C. Kong, D. P. Yu, B. Zhang, W. Fang, and S. Q. Feng, Appl. Phys. Lett. 78, 407 (2000)
29. J. Jie, G. Wang, X. Han, Q. Yu, Y. Liao, G. Li, and J. G. Hou, Chem. Phys. Lett. 387, 466 (2004)
30. X. S. Fang, C. H. Ye, L. D. Zhang, Y. Li, and Z. D. Xiao, Chem. Lett. 34, 436 (2005)
31. R. Yang, Y. Ding, and Z. L. Wang, Nano Lett. 4, 1309 (2004)
32. X. Y. Kong, Y. Ding, R. S. Yang, and Z. L. Wang, Science 303, 1348 (2004)

33. J. Y. Lao, J. G. Wen, and Z. F. Ren, Nano Lett. 2, 1287 (2002)
34. J. Zhang, Y. D. Yang, F. H. Jiang, B. L. Xu, and J. P. Li, J. Solid State Chem. 178, 2804 (2005)
35. L. Xu, Y. Su, Y. Chen, H. Xiao, L. A. Zhu, Q. Zhou, and S. Li, J. Phys. Chem. B 110, 6637 (2006)
36. J. Jie, G. Wang, X. Han, Q. Yu, Y. Liao, G. Li, and J. G. Hou, Chem. Phys. Lett. 387, 466 (2004)
37. Q. Wei, G. W. Meng, X. H. An, Y. F. Hao, and L. D. Zhang, Nanotechnology 16, 2561 (2005)
38. A. Wander, F. Schedin, P. Steadman, A. Norris, R. McGrath, T. S. Turner, G. Thornton, and N. M. Harrison, Phys. Rev. Lett. 86, 3811 (2001)
39. B. Meyer and D. Marx, Phys. Rev. B 67, 035403 (2003)
40. H. J. Fan, B. Fuhrmann, R. Scholz, C. Himcinschi, A. Berger, H. Leipner, A. Dadgar, A. Krost, S. Christiansen, U. Gösele, and M. Zacharias, Nanotechnology 17, S231 (2006)
41. H. J. Fan, B. Fuhrmann, R. Scholz, F. Syrowatka, A. Dadgar, A. Krost, M. Zacharias, J. Crystal Growth 287, 34 (2006)
42. H. J. Fan, W. Lee, R. Hauschild, M. Alexe, G. Le Rhun, R. Scholz, A.Dadgar, K. Nielsch, H. Kalt, A. Krost, M. Zacharias, and U. Gösele, Small 2, 561 (2006)
43. H. J. Fan, F. Fleischer, W. Lee, K. Nielsch, R. Scholz, M. Zacharias, U. Gösele, A. Dadgar, and A. Krost, Superlatt. Microstructures 36, 95 (2004)
44. Y. Ding, X. Y. Kong, and Z. L. Wang, Phys. Rev. B 70, 235408 (2004)
45. K. Zou, X. Y. Qi, X. F. Duan, S. M. Zhou, and X. H. Zhang, Appl. Phys. Lett. 86, 013103 (2005)
46. B. Cheng, Y. Xiao, G. Wu, and L. Zhang, Adv. Funct. Mater. 14, 913 (2004)
47. H. J. Fan, R. Scholz, F. M. Kolb, and M. Zacharias, Appl. Phys. Lett. 85, 4142 (2004)
48. H.J. Fan, R. Scholz, F.M. Kolb, M. Zacharias, U. Gösele, F. Heyroth, C. Eisenschmidt, T. Hempel, and J. Christen，Appl. Phys. A 79, 1895 (2004)
49. L. Manna, D. J. Milliron, A. Meisel, S. C. Scher, and A. P. Alivisatos, Nature Mater. 2, 382 (2003)
50. C. Y. Yeh, Z. W. Lu, S. Froyen, and A. Zunger, Phys. Rev. B 46, 10086 (1992)

51. P. Nguyen, H. T. Ng, T. Yamada, N. K. Smith, J. Li, J. Han, and M. Meyyappan, Nano Lett. 4, 651 (2004)
52. S. Kar, S. Biswas, and S. Chaudhuri, Nanotechnology 16, 3074 (2005)
53. H. J. Fan, A. S. Barnard, M. Zacharias, Appl. Phys. Lett. 90, (2007)
54. A. S. Barnard, J. Phys. Chem. B 110, 24498 (2006)
55. A. S. Barnard, Y. Xiao, and Z. Cai, Chem. Phys. Lett. 419, 313 (2006)
56. A. S. Barnard, X. M. Lin, and L. A. Curtiss, J. Phys. Chem. B 109, 24465 (2005)
57. A. S. Barnard and P. Zapol, J. Chem. Phys. 121, 4276 (2004)
58. A. Wander and N. M. Harrison, Surf. Sci. Lett. 468, 851 (2000)
59. J. A. Kerr, in *CRC Handbook of Chemistry and Physics 1999-2000: A Ready-Reference Book of Chemical and Physical Data*, 81st edition, edited by D. R. Lide (CRC Press, Boca Raton, Florida, 2000), Vol. 9, p. 87.
60. NIST Chemistry Webbook (http://webbook.nist.gov/chemistry)

10

FIELD EMISSION PROPERTIES OF 1-D SiC NANOSTRUCTURES

Guozhen Shen and Di Chen

The synthesis and field emission properties of 1-D SiC nanostructures, including bamboo-like nanowires and hierarchical nanoarchitectures, is described. These nanostructures were grown by a thermochemical method in a vertical tube furnace. SEM and TEM investigations of the structure and morphology of the resulting bamboo-like β-SiC nanowires are described as are electrical measurements of their field emission properties.

10.1 Introduction

The unconventional properties of nanoscale materials have received much attention due to their potential uses in both mesoscopic research and the development of nanodevices. It is well known that nanostructures with sharp tips are promising materials for applications as cold cathode field emission devices. Various materials containing carbon have evoked great interest as cold cathode materials for field electron emission for flat panel displays.[1-3] It is found that field emission can occur at a much lower turn-on field with carbon materials as compared to other materials. The carbon-containing materials include carbon nanotubes, amorphous carbon, diamonds and diamond-like films, and silicon carbide layers.

Silicon carbide (SiC) is a material of great technological interest. There are more than a hundred different crystalline modifications for SiC. The common-occurring polymorphs are 3C, 2H, 4H, and 6H. The electronic structures of these four polytypes are rather different. Zinc blend SiC (3C) has the smallest indirect energy gap of 2.390 eV, whereas the hexagonal polytypes exhibit larger gaps of 3.023 eV (6H), 3.265 eV (4H), and 3.330 eV (2H). Monte Carlo simulation showed the electron mobility along the c-axis direction (the layer stacking direction) of 2H–SiC is higher than that in the 4H– and 6H–SiC. Due to excellent electronic and mechanical properties such as wide band gap, high thermal conductivity, high mechanical strength, high breakdown electric field, and extreme chemical inertness, SiC has attracted intensive research efforts as a potential material for the cold cathode field emission display.[4,5]

This chapter will focus on our recent research activities on the fabrication and field emission properties of 1-D SiC nanostructures. Firstly, a brief survey of the main synthetic procedures for 1-D SiC nanostructures is given. The following part will then discuss the SiC nanostructures fabricated in our group. It mainly focuses on the unconventional bamboo-like SiC nanowires and the hierarchical SiC nanoarchitectures. And then the field emission properties of these 1-D SiC nanostructures will be discussed in detail. Finally, we will give our own perspectives.

10.2 Synthetic Methods for 1-D SiC Nanostructures

Since the first synthesis of SiC nanorods in 1995 by using carbon nanotubes as a template, several techniques have been developed to fabricate 1-D SiC nanostructures, such as the carbon nanotube confined method, chemical vapor deposition methods, and solution methods. The following part will give a brief survey of these methodologies.

10.2.1 Carbon nanotube confined method

The carbon nanotubes confined method is very promising since during reaction the very stable carbon nanotubes act as a template, spatially confining the reaction to the nanotubes, which results in the formation of nanorods with similar diameter and length to those of the carbon nanotubes. SiC nanowires were first synthesized by reacting carbon nanotubes with SiO or Si–I_2 at about 1200°C.[6] Using a two-step reaction, in which SiO vapor was first generated via the reduction of silica and then reacted with carbon nanotubes,[7] Han et al. synthesized SiC nanorods at 1400°C. Based on a similar method, vertically aligned SiC nanowires have been synthesized by reacting aligned carbon nanotubes with SiO.[8] The SiC nanowires in the arrays have similar diameters (10–40 nm), spacings (~100 nm), and lengths (up to 2 mm). The SiC nanowire possesses a high density of planar defects and stacking faults which are perpendicular to the wire axes.

10.2.2 Chemical Vapor Deposition methods

Chemical vapor deposition (CVD) methods are the most commonly used techniques for the synthesis of 1-D SiC nanostructures. There are many reports on the synthesis of 1-D SiC nanostructures using CVD methods. For example, in a hot filament CVD process, SiC nanorods can be produced on silicon using a solid carbon and silicon source.[9] SiC nanorods with diameters of 5–20 nm and lengths of about 1 μm have been grown on porous silicon substrates by CVD using iron as the catalyst.[10] A tablet composed of pressed Si and SiO_2 powders is placed in the chamber in which the reaction takes place in the presence of CH_4 and H_2.[11] β-SiC nanorods can also be synthesized by using a solid mixture of graphite, silicon, and silicon dioxide. Silicon carbide nanowires with diameters of 20–50 nm and lengths of several micrometers have been fabricated on Ni (with a thickness of 2 nm) deposited on silicon substrates by CVD.[12] Amorphous silicon carbide nanosprings as well as biphase (crystalline core/amorphous sheath) helical nanowires have recently been synthesized by plasma enhanced CVD.[13]

10.2.3 Solution methods

Recently, direct synthesis of 1-D SiC nanostructures in solution at relatively low temperature have been reported and received great research attention. For instance, SiC nanorods are synthesized by a single-step reaction at 400°C in an autoclave with $SiCl_4$ and CCl_4 as the reactants and metallic Na as co-reductant.[14] The reaction can be described as: $SiCl_4 + CCl_4 + 8Na \rightarrow SiC + 8NaCl$. The SiC nanorods thus produced have diameters ranging from 10 to 40 nm and lengths up to several micrometers. When CCl_4 was substituted by C_6Cl_6, SiC nanowires and C/SiC nanocables could be produced by tuning the experimental parameters.[15] Using $SiCl_4$, 2-ethoxyethanol ($HOCH_2CH_2OCH_2CH_3$) as the source materials and Mg ribbons as the catalysts, aligned SiC nanorods and nanoneedles were synthesized in stainless steel autoclave at 600°C.[16] By a reaction of CH_3CH_2OH, $SiCl_4$, and Li in an autoclave at 600°C, SiC nanobelts have been prepared, which are usually 50-200 nm wide, 20-60 nm thick and up to tens of micrometers long.[17]

10.3 Bamboo-like and Hierarchical 1-D SiC Nanostructures

The fabrication of complex nanoarchitectures with controlled morphology, orientation and dimensionality have attracted significant attention over the last decade since such control is crucial for the determination of structure-property relationships in many processes, development of new pathways for materials synthesis and novel applications of nanostructured materials. It is important to synthesize 1-D SiC nanostructures with special morphologies to fulfill their usage potential in practical applications.

We focused on the synthesis of special 1-D SiC nanostructures using thermochemical process.[18-20] In this process, a vertical induction furnace, which consisted of a fused quartz tube and an induction-heated cylinder made of high purity graphite coated with a C fiber thermo insulating layer, which had inlet and outlet C pipes on its top and base, respectively, were used as the reaction system. This system provides more rapid temperature change than the conventional tube furnace, which favors the formation of special nanostructures. Bamboo-like SiC nanowires were synthesized using SiO, GaN and graphite powder as the source materials. The products obtained consist of interesting bamboo-like structures on a very large scale as shown in Figure 10.1a and 10.1b. SEM images in Figure 10.1c-i show a series of typical bamboo-like SiC nanowires obtained. A typical bamboo-like SiC nanowire has hexagonal cross-section, 80-300 nm in diameter, up to several hundred microns in length. Governed by the vapor-liquid-solid (VLS) mechanism, all the bamboo-like SiC nanowires have one plate tip and one particle-attached tip as shown in these images. Contrary to the conventional SiC nanowires that have the same diameter along the whole length, the bamboo-like SiC nanowires have clearly waved diameters. Bigger

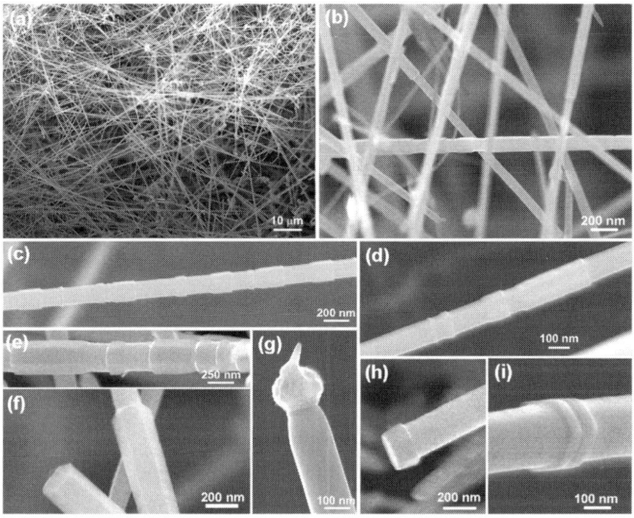

Fig 10.1 SEM images of the bamboo-like β-SiC nanowires. a) Low-magnification view; b) High-magnification view; c-i) High-magnification view, which clearly indicate the formation of bamboo-like nanowires with hexagonal cross-sections. (g) and (h) clearly show that the SiC nanowire has one flat tip and one catalyst-attached tip.

diameter knots grew almost periodically along the whole length of the bamboo-like nanowires, which look more like bowls wrapping on the nanowires, as shown in these images.

The whole bamboo-like SiC nanowire has the same growth direction; along a universal <111> direction, though the stem and the knots have quite different diameters, which can be verified by selected area electron diffraction (SAED) patterns taken from the corresponding parts. Figures 10.2b and 10.2c are the SAED patterns taken from the area marked with circles in Figure 10.2a, which correspond to the stem and knot, respectively. They clearly show the same orientation. The SAED pattern from the stem shows only bright spots while that from the knot shows bright spots as well as streaks, indicating that defects only exist in the knot area. It is well known that since the electron diffraction pattern streaks are always perpendicular to the stacking fault, the existence of (111) stacking faults is confirmed by aligning the electron beam along the [110] direction of β-SiC nanowires. High-resolution transmission electron microscopy (HRTEM) investigations (Figure 10.2d, 10.2e)

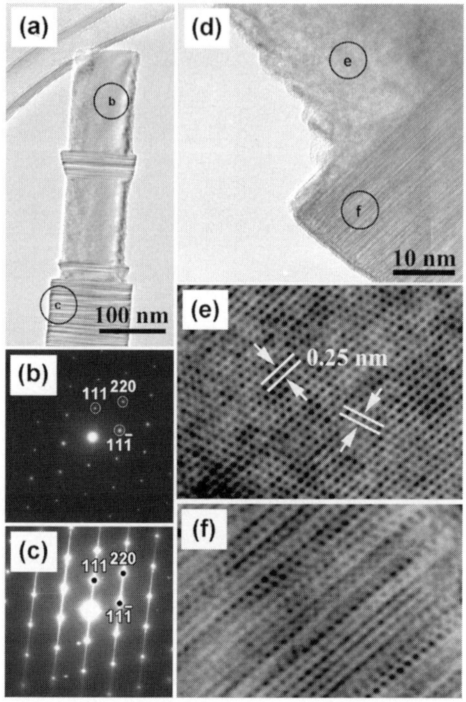

Fig 10.2 a) TEM image of a single bamboo-like β-SiC nanowire; b,e) SAED pattern taken from the bamboo-like β-SiC nanowire marked in a, respectively; d) TEM image of the joint part of the stem and a knot; e) HRTEM lattice image taken from the stem; f) HRTEM lattice image taken from the knot.

confirm that the stem of the bamboo-like SiC nanowire is structurally uniform and no stacking faults or other planar defects are observed within it, while for the knot of nanowire, stacking faults exist in the whole knot.

The growth of bamboo-like SiC nanowires can be explained as follows. At high reaction temperature, Ga is formed due to the decomposition of GaN and the newly formed Ga may be in the form of small-sized liquid clusters. The Ga clusters are then transported by the carrier gas to a lower temperature region, where they deposit in the form of liquid droplets on the inner wall of the graphite crucible. These droplets are favorable sites for the further growth of SiC nanostructures. At the same time, solid SiO is evaporated at high temperature to generate

SiO gas. And the SiO gas reacts with graphite powder to form SiC clusters. Governed by the VLS mechanism, the SiC clusters would continuously precipitate out from the liquid Ga droplets, allowing the storage of the elastic energy in the small volume of the solid SiC phase. With the continuous precipitation of SiC phase, the stored elastic energy will be released by producing stacking faults, which will then result in the formation of knots along the SiC nanowires.

Based on a similar VLS mechanism, novel hierarchical SiC nanoarchitectures, which look like strings made of platelet structures were produced by thermal evaporation of SiO, Ga_2O_3 and graphite powders with controlled gas protection.[20] The string-like structures have diameters of 50-70 nm and lengths of tens of micrometers. Figure 10.3a shows the TEM image of the synthesized product, which are hierarchical in structure. It can be seen that all the branched nanoplatelets possess a well-preserved orientation relationship. All the nanoplatelets grow out of the stems and are parallel to each other. The nanoplatelets have uniform diameters of ~ 5-10 nm with a large plate-size distribution ranging from 10 to 150 nm. In fact, it was thought that the lack of evenness in the plate-size was due to the high reaction temperature and rapid growth rate

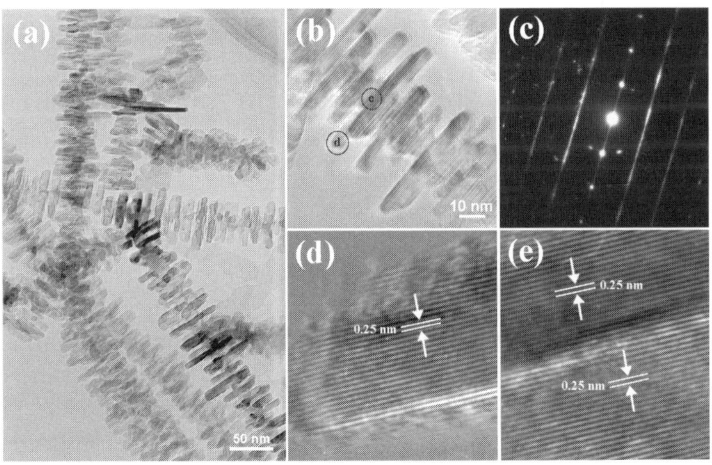

Fig 10.3 (a) TEM image of the hierarchical SiC nanoarchitectures. (b) TEM image of a single hierarchical SiC nanoarchitecture. (c) Corresponding SAED pattern. (d) and (e) HRTEM images recorded from the parts marked in Figure 10.3b, respectively.

in the induction furnace system. Figure 10.3b is a TEM image, clearly showing the hierarchical structure of the product. Investigations by X-ray energy dispersive spectrometer (EDS) reveal that both the stem and branched platelets are pure SiC phase. A SAED pattern of the hierarchical structure shown in Figure 10.3b is depicted in Figure 10.3c, which is recorded along the [110] zone axis. Bright spots as well as streaks are observed in this pattern, indicating that some defects exist within the structure, which is quite similar to the synthesized bamboo-like SiC nanowires.

10.4 Field Emission Properties of 1-D SiC Nanostructures

The emission characteristics of field emission devices (FEDs) strongly depend on the device geometry such as tip sharpness R and height h, anode–cathode distance, emitter density, and material, etc.

SiC is superior as a cold cathode material for its wide band gap, high thermal conductivity, high mechanical strength, high breakdown electric field, and extreme chemical inertness compared with carbon nanotubes. Field emission properties of different 1-D nanostructures have been extensively studied. Vertically-aligned SiC nanowires synthesized from carbon nanotubes show excellent field emission properties.[8] The observed electron emission turn-on field (E_{to}) and threshold field (E_{thr}), defined as the macroscopic fields required to produce a current density of 10 μA/cm^2 and 10 mA/cm^2, respectively, are about 0.9 V/μm and 2.7 V/μm. The emission behavior is attributed to the very high density of emitting tips with a small ratio of curvature at the emitting surface of the SiC nanowires.

It is well known that the field emission properties are greatly influenced by the density and tip structures of the nanostructures studied. It is important to get different kinds of nanostructures with different morphologies to investigate their field emission properties. Taking 1-D ZnO nanostructures as an example can explain such influences. It was found that vertically aligned ZnO nanoneedles with sharp tips have a turn-on field of about 6.2 V/μm, while vertically aligned ZnO nanopencils and nanonails have turn-on field of 7.2 V/μm and 7.9 V/μm, respectively. These kinds of ZnO nanostructures have obviously different tips and thus resulted in the difference in field emission properties.[21,22] Figure 10.4 shows the emission current density, *J*, versus applied field curves at an anode-sample distance of 100 μm of the bamboo-like SiC

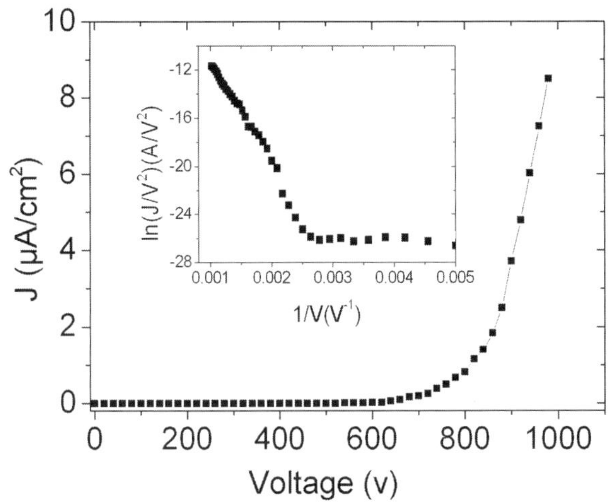

Fig 10.4 Field-emission J-V curves of the bamboo-like β-SiC nnowires. The inset is the corresponding Fowler-Nordheim (FN) plots.

nanowires. The observed turn-on field was about 10.1 V/μm. Though the value is higher than the vertically aligned SiC nanowires,[8] it is much lower than the randomly distributed SiC nanowires with uniform diameters along the whole length with round cross-sections.[9,10,23] In fact, compared with the conventional SiC nanowires which have round cross-sections, bamboo-like SiC nanowires have highly faceted hexagonal cross-sections and the sharp corners of the nanowires act as efficient electron emitting sites. Besides, high density stacking faults exist in the bamboo-like SiC nanowires. These allow electron emission at a lower voltage from bamboo-like nanowires by increasing the field enhancement factor β, which shows strong dependence on the geometry of materials.

The fabrication of complex nanoarchitectures with controlled morphology, orientation and dimensionality is also very important. These complex structures may show unique properties compared with simple ones. Figure 10.5 shows the field-emission J-V curves recorded from the synthesized hierarchical SiC nanoarchitectures. A turn-on field of 12 V/μm was obtained for these complex SiC structures.

The Fowler-Nordheim plot of the nanostructures was usually investigated besides the Field-emission J-V curves.[24] The Fowler-Nordheim law relates the emitted current and the local field E at the

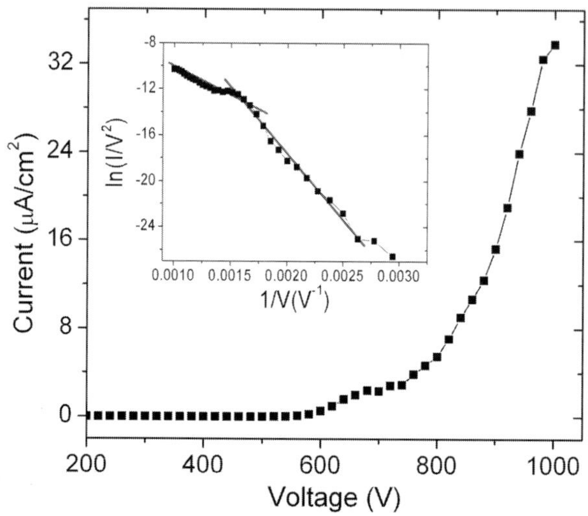

Fig 10.5 Field-emission J-V curves recorded from hierarchical SiC nanoarchitectures. The inset is the corresponding Fowler-Nordheim (FN) plot in which two distinct linear regimes are highlighted with the red lines.

emitter surface,

$$J - (E^2\beta^2/\Phi)\exp(-B\Phi^{3/2}/E\beta) \quad (?)$$

Where J is the emission-current density, E is the local applied field (where $E=Vd^{-1}$, and d is the anode-sample distance), Φ is the work function of the material, β is the enhancement factor, and B is a constant. According to the classical FN theory, the β factor is defined as the ratio of the local surface electric field of the vacuum gap. It is a token of the electric field enhancement effect due to the geometry structure of the emitters, thus the FN plot should have a linear behavior. However, in our case for both the bamboo-like SiC nanowires and the hierarchical SiC nanoarchitectures, two linear behaviors in the measurement range were observed instead. It was thought that these kinds of behaviors were caused by current saturation, which is attributed to possible adsorbents on the nanostructure appearing under a higher electric field or caused by space charge effects.

In fact, till now, the reported turn-on fields of 1-D SiC nanostructures varied in a large range of about 1 to 27 V/μm. The values are greatly influenced by the microstructures of the 1-D SiC nanostructures. Though many kinds of 1-D SiC nanostructures have been fabricated, it is still a change to control the microstructures of SiC nanostructures to get better field emission properties to enable real world applications of these nanostructures.

Though it is well known that refractory bulk SiC exhibits excellent oxidation resistance at high temperature, investigations have found that 1-D SiC nanostructures exhibit strong oxidization due to their large specific surface area. The oxidation interaction between the inner SiC nanostructures and the oxide layers has great influence on their practical usage in the field of electronics. The SiO_2 layers covered on SiC nanowires, to some content, damaged the excellent field emission properties of SiC materials. SiC nanowires coated with BN were synthesized by the thermochemical process. BN was used as the coating material is due to its unique chemical and physical properties, such as low density, high melting point, and chemical inertness. Field emission results show that the turn-on field decreased from 13 V/μm for the pure SiC nanowires to about 6 V/μm for the BN-coated SiC nanowires.[25] That means the field emission properties of BN-coated SiC nanowires have been greatly improved compared with that of the pure SiC nanowires.

10.5 Conclusions

In this chapter, we reviewed the synthesis and field emission properties of 1-D SiC nanostructures, including bamboo-like nanowires and hierarchical nanoarchitectures, which have been developed in our group by the thermochemical method. It was found that the field emission properties varied for different kinds of nanostructures.

The investigation of 1-D SiC nanostructures will remain exciting and highly rewarding. There is much innovation and development waiting to happen and a lot of issues to be investigated. These include:

(1) Precise controlled growth of 1-D SiC nanostructures in their position, alignment, density, morphology is required in order to ensure better field emission properties. There is much to be investigated in the theoretical understanding of the effect of nanostructures on field emission properties.

(2) It is important, and is still a challenge, to find suitable coating materials, besides BN, to avoid surface oxidation and to improve the field emission properties of 1-D SiC nanostructures.

(3) Finally, practical field emitters based on 1-D SiC nanostructures need to be developed.

References

1. Q. H. Wang, A. A. Setlur, J. M. Lauerhaas, J. Y. Dai, E. W. Seelig and R. P. H. Chang, *A nanotube-based field-emission flat panel display*, Appl. Phys. Lett. 72, 2912 (1998)
2. K. Okano, K. Hoshina, M. Iida, S. Koizumi and T. Inuzuka, *Fabrication of a diamond field emitter array*, Appl. Phys. Lett. 64, 2742 (1994).
3. P. W. May, S. Hohn, M. N. R. Ashfold, W. N. Wang, N. A. Fox, T. J. Davis and J. W. Steeds, *Field emission from chemical vapor deposited diamond and diamond-like carbon films: Investigations of surface damage and conduction mechanisms*, J. Appl. Phys. 84, 1618 (1998)
4. S. Z. Deng, Z. B. Li, W. L. Wang, N. S. Xu, J. Zhou, X. G. Zheng, H. T. Xu, J. Chen and J. C. She, *Field emission study of SiC nanowires/nanorods directly grown on SiC ceramic substrate*, Appl. Phys. Lett. 89, 023105 (2006)
5. J. Nishida, *Field emission from SiC whisker*, J. Appl. Phys. 38, 5417 (1967)
6. H. J. Dai, E. W. Wong, Y. Z. Lu, S. Fan and C. M. Lieber, *Synthesis and characterization of carbide nanorods*, Nature 375, 769 (1995)
7. W. Q. Han, S. Fan, Q. Lu, W. J. Liang, B. L. Gu and D. P. Yu, *Continuous synthesis and characterization of silicon carbide nanorods*, Chem. Phys. Lett. 265, 374 (1997)
8. Z. Pan, H. L. Lai, F. C. K. Au, X. Duan, W. Zhou, W. Shi, N. Wang, C. S. Lee, N. B. Wong, S. T. Lee and S. Xie, *Oriented silicon carbide nanowires: synthesis and field emission properties*, Adv. Mater. 12, 1186 (2000).
9. X. T. Zhou, N. Wang, H. L. Lai, H. Y. Peng, I. Bello, N. B. Wong, C. S. Lee and S. T. Lee, *β-SiC nanorods synthesized by hot filament chemical vapor deposition,* Appl. Phys. Lett. 74, 3942 (1999).
10. X. T. Zhou, H. L. Lai, H. Y. Peng, F. C. K. Au, L. S. Liao, N. Wang, I. Bello, C. S. Lee and S. T. Lee, *Thin β-SiC nanorods and their field emission properties,* Chem. Phys. Lett. 318, 58 (2000)
11. H. L. Lai, N. B. Wong, X. T. Zhou, H. Y. Peng, F. C. K. Au, N. Wang, I. Bello, C. S. Lee, S. T. Lee and X. F. Duan, *Straight β-SiC*

nanorods synthesized by using C-Si-SiO_2, Appl. Phys. Lett. 76, 294 (2000)
12. H. J. Choi, H. K. Seong, J. C. Lee and Y. M. Sung, *Growth and modulation of silicon carbide nanowires*, J. Cryst. Growth 269, 472 (2004).
13. D. Zhang, A. Alkhateeb, H. Han, H. Mahmood, D. N. McIlroy and M. G. Norton, *Silicon carbide nanosprings*, Nano Lett. 3, 983 (2003)
14. Q. Lu, J. Q. Hu, K. B. Tang, Y. T. Qian and G. Zhou, *Growth of SiC nanorods at low temperature*, Appl. Phys. Lett. 75, 507 (1999)
15. G. Z. Shen, D. Chen, K. B. Tang, Y. T. Qian and S. Y. Zhang, *Silicon carbide hollow nanospheres, nanowires and coaxial nanowires*, Chem. Phys. Lett. 375, 177 (2003)
16. G. C. Xi, Y. K. Liu, X. Y. Liu, X. Q. Wang and Y. T. Qian, *Mg-catalyzed autoclave synthesis of aligned silicon carbide nanostructures*, J. Phys. Chem. B 110, 14172 (2006)
17. G. C. Xi, Y. Y. Peng, S. M. Wan, T. W. Li, W. C. Yu and Y. T. Qian, *Lithium-assisted synthesis and characterization of crystalline 3C-SiC nanobelts*, J. Phys. Chem. B 108, 20102 (2004).
18. Y. H. Gao, Y. Bando, K. Kurashima and T. Sato, *The microstructural analysis of SiC nanorods synthesized through carbothermal reduction*, Script. Mater. 44, 1941 (2001)
19. G. Z. Shen, Y. Bando, C. H. Ye, B. D. Liu and D. Golberg, *Synthesis, characterization and field emission properties of bamboo-like β-SiC nanowires*, Nanotechnology 17, 3468 (2006)
20. G. Z. Shen, Y. Bando and D. Golberg, *Self-assembled hierarchical single-crystalline β-SiC nanoarchitectures,* Cryst. Growth & Design 7, 35 (2007)
21. H. Ham, G. Z. Shen, J. H. Cho, T. Lee, S. H. Seo and C. J. Lee, *Vertically-aligned ZnO nanowires produced by a catalyst-free thermal evaporation method and their field emission properties*, Chem. Phys. Lett. 404, 69 (2005)
22. G. Z. Shen, Y. Bando, B. D. Liu, D. Golberg and C. J. Lee, *Characterization and field-emission properties of vertically-aligned ZnO nanonails and nanopencils fabricated by a modified thermal evaporation process*, Adv. Funct. Mater. 16, 410 (2006)
23. T. H. Yang, C. H. Chen, A. Chatterjee, H. Y. Li, J. T. Lo, C. T. Wu, K. H. Chen and L. C. Chen, *Controlled growth of silicon carbide nanorods by rapid thermal process and their field emission properties*, Chem. Phys. Lett. 379, 155 (2003)

24. R. H. Fowler and L. Nordheim, *Electron emission in intense electric fields*, Proc. Royal Soc., London A119, 173 (1928)
25. C. C. Tang and Y. Bando, *Effect of BN coatings on oxidation resistance and field emission of SiC nanowires*, Appl. Phys. Lett. 83, 659 (2003).

11

MANIPULATING THE OPTICAL PROPERTIES OF INDIVIDUAL AND ARRAYS OF GOLD NANOPYRAMIDS

Jeunghoon Lee, Joel Henzie and Teri W. Odom

This chapter describes the fabrication and optical characterization of a new type of metal nanoparticle: gold nanopyramids. Individual pyramids exhibit orientation-dependent scattering under unpolarized white light, which can be used to determine their relative orientation on a substrate without direct imaging tools. Arrays of gold pyramids encapsulated within a polymer film exhibit polarization and wavevector-dependent optical properties that match well with calculations. The scattering peaks observed at visible wavelengths can be assigned to multi-polar surface plasmon resonances, which originate from the large size and anisotropic shape of the pyramids. Unique prospects of these three-dimensional metal particles will also be discussed.

11.1 Introduction

Metal nanoparticles exhibit interesting optical, magnetic, and catalytic properties that are scientifically and technologically important[1,2]. For example, the optical properties of metal nanoparticles are interesting because of their broad tunability from ultraviolet to near-infrared wavelengths, which can be exploited in nanoscale photonics[3], chemical and biological sensing[4] and high-efficiency photovoltaic devices[5]. Recent advances in both solution-based syntheses and fabrication techniques to create particles with control over size, shape, and materials have driven the discovery of their physical properties as well as their prospects in emerging applications. Most research efforts on noble metal (Ag, Au, Cu) nanoparticles have focused on chemical synthesis because the preparative techniques are relatively straightforward; the size and shape of nanoparticles can be tailored by controlling conditions such as reaction temperature, surfactants, and concentrations of precursors[1, 6]. Spherical and rod-shaped nanoparticles were among the first to be prepared and thus have been the most intensely studied systems. Metal nanoparticles with more complex shapes, such as cubes[7], polyhedrons[8], cages[9], shells[10] and stars[11] have also been synthesized.

Because the shape of large (>100 nm) metallic particles is often difficult to control by synthetic methods, fabrication techniques have been pursued as an alternative approach. Top-down procedures can fabricate metal structures with anisotropic shapes and with optical properties tunable from 500 to 3000 nm[4, 12]. Serial patterning techniques such as electron beam lithography can generate structures with arbitrary sizes and shapes in 2D[13, 14], although low throughput and small write-areas (hundreds of square microns) are current challenges. Besides direct-writing methods, templates have also been used to produce particles with anisotropic shapes. Isolated sub-micron spheres and close-packed sphere arrays have acted as templates onto which metals were deposited to obtain particles with crescent-like[15, 16] and triangular shapes[17], respectively. Anodized aluminum oxide membranes are another widely used template for producing nanorods as long as 10 μm[18]. Recently, we developed a template-based approach to fabricate metallic pyramidal shells with smooth facets and sharp tips ($r < 2$ nm)[19].

11.2 Optical Properties of Metal Nanostructures

Light interacts with metal surfaces through the resonant excitation of surface free electrons to produce collective electron density oscillations called surface plasmons (SPs). When confined to a finite volume, such as in the case of metallic nanoparticles, the SPs are referred to as localized surface plasmons (LSPs). These resonances are extremely sensitive to the shape[1], size[20] and dielectric environment[21] of the nanoparticles. Spherical particles with sizes less than 50 nm support single LSP resonances that are dipolar in character. Accordingly, their optical properties can be described reasonably well by the lowest order term in Mie theory[22]. Also, 100-nm diameter Au pyramids show only a dipole resonance in contrast to pyramids with sizes of 250 nm or more[23]. Rod-shaped Au nanoparticles exhibit both transverse and longitudinal modes depending on polarization[6, 24].

Larger metallic particles (overall sizes > 100 nm) with anisotropic shapes can exhibit multiple LSP resonances[25, 26] that correspond to higher order plasmon modes[27]. These multipolar excitations depend both on the direction of the wavevector as well as the polarization vector[28]. Thus, certain excitation angles can allow selected resonances to be more pronounced[29]. Disordered assemblies of 100-nm Ag particles embedded in poly(dimethylsiloxane) (PDMS) films exhibited only a dipole resonance, but a quadrupole resonance emerged as the film was stretched in 2D[30]. In addition, ordered arrays of 250-nm Au pyramids embedded in PDMS displayed multipolar plasmon resonances that depended strongly on the direction of the wavevector[23].

Fabrication and Spectroscopy of Nanopyramids

As discussed in the previous section, most studies on the optical properties of metallic nanoparticles have focused on particles whose shapes were symmetric. By developing a simple nanoscale fabrication technique called PEEL, we have developed a way to produce metal pyramids that are relatively monodisperse (95%), can be made out of different materials, and exhibit unique optical properties.

11.3 PEEL: Fabrication Methodology for Nanopyramids

PEEL is a procedure consisting of four major steps to fabricate pyramidal nanoparticles: **P**hase-shifting photolithography, **E**tching, **E**lectron-beam deposition, and **L**ift-off[19, 31]. Figure 11.1 highlights the key, intermediate stages in the technique. First, an array of photoresist posts (100-250 nm diameters) was generated by exposing photoresist (Shipley 1805) through a *h*-PDMS mask patterned with an array of cylinders in bas-relief and then developing the resist (Fig. 11.1A). Next,

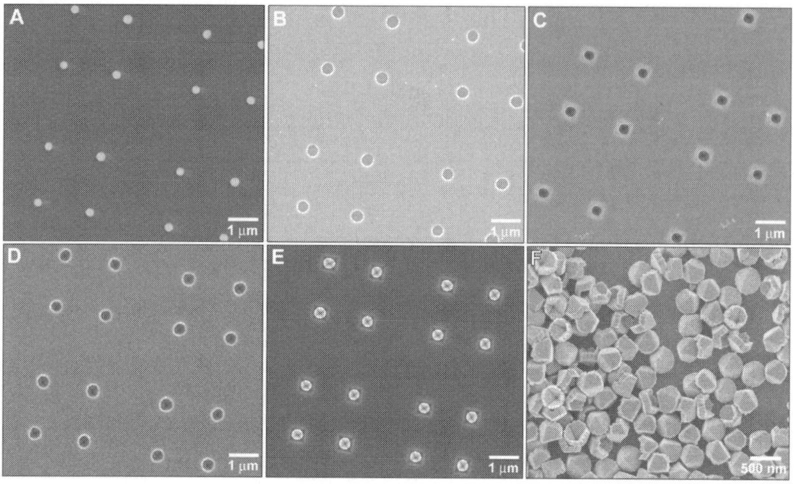

Figure 11.1 Scanning electron microscopy (SEM) images of the key fabrications steps. **(A)** Array of photoresist posts patterned with phase-shifting photolithography. **(B)** Cr holes after Cr deposition and removal of photoresist. **(C)** Pyramidal pits formed after an anisotropic Si etch. **(D)** Gold deposited inside the pits and on Cr film. **(E)** Gold nanopyramids embedded in Si substrate after peeling off the gold film. **(F)** Gold nanopyramids dispersed on a substrate.

a thin layer (~ 20 nm) of Cr was deposited and lift-off was performed to produce round holes with smooth edges—features that were critical for obtaining nanopyramids with sharp tips (Fig. 11.1B). A subsequent anisotropic Si etch (potassium hydroxide/isopropanol) created pyramidal pits beneath each Cr hole (Fig. 11.1C). Metal (usually Au) was then deposited by e-beam to a desired thickness (t) (Fig. 11.1D). Finally, the Cr layer was etched to peel off the Au film perforated with arrays of holes and also to reveal Au nanopyramids embedded in the Si template (Fig. 11.1E). The pyramids were then either released by sonication and dispersed on a surface (Fig. 11.1F) or encapsulated as ordered arrays in PDMS (Fig. 11.2).

11.4 Encapsulation of Arrays of Nanopyramids

Figure 11.2 outlines a procedure for transferring Au nanopyramids situated within the etched pits of a Si (100) template into a transparent PDMS film. First, the Si substrate containing the pyramids (Fig. 11.2A)

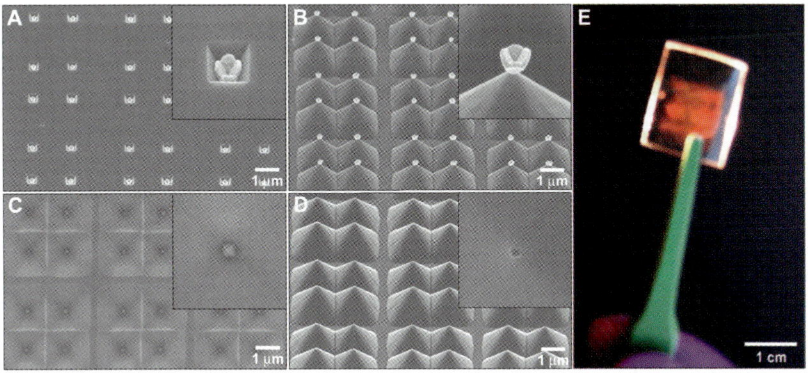

Figure 11.2 SEM images of the steps to encapsulate arrays of pyramids in a uniform dielectric environment. **(A)** 250-nm diameter Au pyramids situated in the centers of etched Si pits. **(B)** Etched Si (100) pedestals supporting the Au pyramids. **(C)** Gold pyramids transferred and partially embedded within the PDMS film. **(D)** Etched Si (100) pedestals after removal of the pyramids. The dimensions of all insets are 1 μm x 1 μm.

was subjected to an aniosotropic Si etch. The nanopyramids protected the underlying Si, and thus the exposed Si (100) surface was etched quickly to result in Au pyramids supported on Si pedestals (Fig. 11.2B). To improve the adhesion between the pyramids and the PDMS matrix, the Si substrate was first passivated with tridecafluoro-1,1,2,2-tetrahydrooctyl-1-trichlorosilane, and the Au pyramids were functionalized with (3-mercaptopropyl) trimethoxysilane.[32] The sample was then pressed against a thin (10 μm) layer of unpolymerized *h*-PDMS[33] and cured. The *h*-PDMS film (with the pyramids now partially embedded) was pulled off the substrate with tweezers, and the pyramidal tips protruded partially from the PDMS mold (Fig. 11.2C, inset). We checked that the pyramids were transferred into PDMS by characterizing the etched Si substrate (Fig. 11.2D). To encapsulate the gold pyramids fully, we exposed the array of pyramidal tips to mercaptosilane and then spin-coated another thin (10 μm) layer of *h*-PDMS on top. Fig. 11.2E is a photograph of a large-area PDMS film encapsulating a 1-cm^2 array of 250-nm Au pyramids.

11.5 Optical Characterization of Gold Nanopyramids

An inverted optical microscope (Nikon TE-2000U) equipped with a dark field (DF) condenser (numerical aperture (NA) = 0.8 – 0.95) was used to

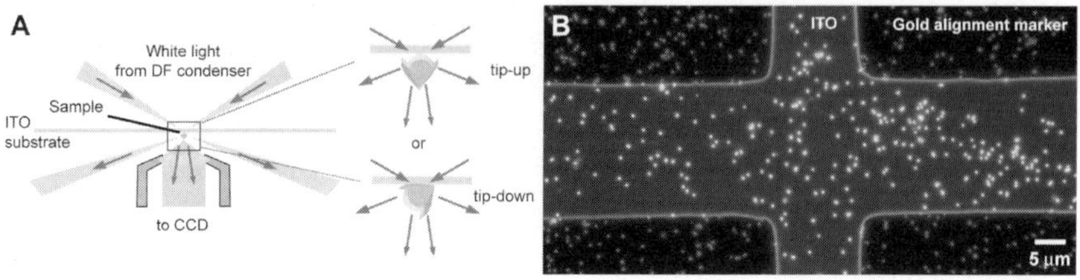

Figure 11.3 (A) Schematic dark field (DF) microscopy set-up. **(B)** DF scattering image of well-separated 300-nm Au pyramids on ITO-coated glass substrates with alignment markers.

image light scattered by the Au nanopyramids. In DF spectroscopy, light is passed through a condenser and is incident on the sample at very high angles. Thus, only light scattered by the sample (in this case, nanopyramids) is collected through an objective lens (Fig. 11.3A) and sent to a liquid-N_2 cooled CCD detector (Jobin Yvon Symphony, 1024 × 256 pixels) and spectrometer (Jobin Yvon Triax 552). The CCD imaging detector allowed us to carry out single particle measurements by selecting specific pixel ranges on the array. Background scattering from the transparent substrate was subtracted from the raw spectrum containing spectral information from both the pyramids and the substrate and was then normalized against the spectral profile of the halogen light source[34].

To correlate the size and shape of the nanopyramids with their spectral properties, nanopyramids dispersed in water were drop-cast onto indium tin oxide (ITO) substrates patterned with Au alignment markers (Fig. 11.3B). These substrates were prepared by evaporating a thin (10 nm) layer of Au through a Cu TEM grid[11].

Optical Properties of Individual Gold Nanopyramids

11.6 Orientation-dependent Scattering Properties

Gold nanopyramids dispersed on a flat surface typically exhibited two orientations: either their tips pointed up (*away* from the surface) or pointed down (*touched* the surface) (Fig. 11.3A). When viewed in a DF optical image, single pyramids appeared as uniform, round spots because of the diffraction limit of the microscope (Fig. 11.3B). To correlate the scattering spectra of a single nanopyramid with its orientation, both optical measurements and SEM imaging were carried out on 300-nm Au pyramids with different thicknesses (t = 60, 40, and 15 nm). The spectra of thicker nanopyramids (t = 40 and 60 nm) showed a more pronounced dependence on their orientation relative to the direction of incident light (Fig. 11.4). Noticeably, the scattering spectrum of a 60-nm thick nanopyramid whose tip pointed up was dominated by strong scattering at 600 nm, while those with tips pointed down showed two major peaks at 600 nm and 850 nm with comparable intensity (Fig. 11.4A). The 600-nm peak in the scattering spectrum of a 40-nm thick nanopyramid was not as dominant as the 60-nm thick case, but the same trend could be observed for the different orientations of the pyramids (Fig. 11.4B).

Discrete dipole approximation (DDA) calculations of nanopyramids with a 300-nm diameter base, performed in collaboration with the Schatz group at Northwestern University[23], indicated that the 850-nm peak can

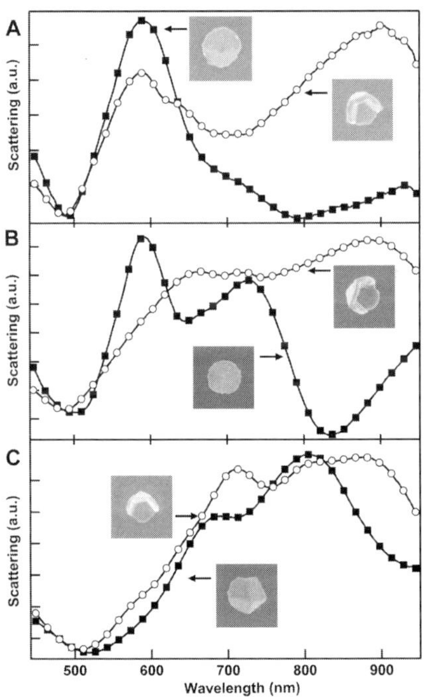

Figure 11.4 Single particle scattering spectra of 300-nm Au pyramids with thicknesses of **(A)** 60 nm, **(B)** 40 nm, and **(C)** 15 nm. Spectra with tips pointing up and pointing down were distinctly different. All SEM insets are 600 nm × 600 nm.

be assigned to a quadrupole plasmon resonance. Thicker nanopyramids scattered light with a higher efficiency at 600 nm, indicating the surface plasmon resonance mode responsible for the 600-nm peak was enhanced when the distance between the inner and outer faces of the pyramids was increased. When the tip of a nanopyramid points up, the incident light impinges on the edge of the nanopyramid base (Fig. 11.3A) and could selectively excite plasmon resonance modes that depend on the thickness of the pyramids. As expected, such resonance modes were absent in the scattering spectra of the thinner (15 nm) pyramids (Fig. 11.4C). These distinct orientation-dependent optical properties of the Au pyramids under *unpolarized* light are only observed because of the unique shape of the pyramidal shells.

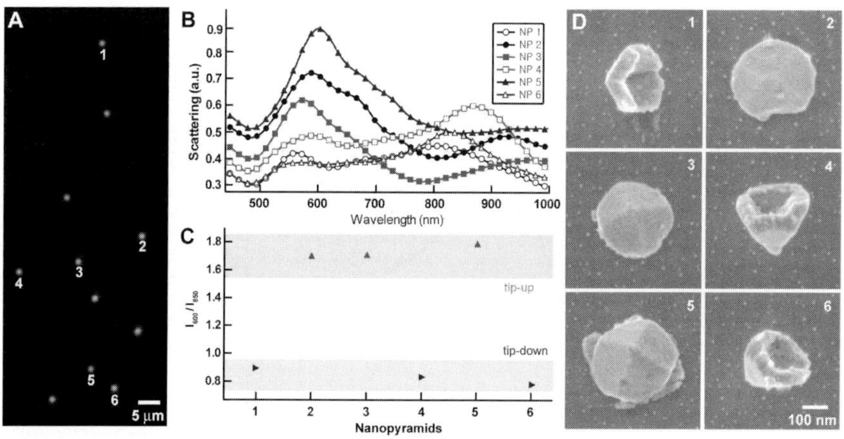

Figure 11.5 (A) DF micrograph of isolated Au pyramids deposited on ITO-coated glass. **(B)** Scattering spectra of six individual Au pyramids indicated in **(A)**. **(C)** Ratio of normalized scattering intensity at 600 nm and 850 nm for pyramids corresponding to those indicated in **(A)**. **(D)** SEM images of the six pyramids, which confirmed that their relative orientation could be determined only by their optical properties.

11.7 Spectral Determination of the Orientation of Gold Nanopyramids

To test whether such spectral differences could be used to determine the relative orientations of the nanopyramids on a surface, we dispersed dilute solutions of 300-nm Au pyramids (t = 70 nm) onto ITO-coated glass substrates. We then selected at random six spots in the DF optical image (Fig. 11.5A), each corresponding to a single pyramid, and measured their scattering spectra (Fig. 11.5B). This spectral information was then collected and displayed as a ratio of the normalized scattering intensity at 600 nm to that at 850 nm, which separated into two distinct bands (Fig. 11.5C). Based on the trends observed in Fig. 11.4, the scattering spots labeled 2, 3, and 5 should be pyramids whose tips point up. As expected, SEM images and the predicted nanopyramid orientation matched perfectly (Fig. 11.5D).

Optical Properties of Arrays of Gold Nanopyramids

11.8 Scattering Properties of Arrays of Pyramids with Tips Pointed Perpendicular to a Surface (Orientation I)

Since the orientations of randomly dispersed nanopyramids on a substrate are limited to either tips pointing away (and up) or tips pointing toward (and touching down on) the surface, we needed to devise a different strategy to manipulate the orientation of the pyramids. This improved level of control over orientation is important to correlate the orientation of the Au pyramids with *specific* plasmon modes directly. To fix the pyramids at well-defined angles relative to a substrate, we encapsulated them in thin films of *h*-PDMS, and then sliced the film at different angles relative to the plane of the array.[31]

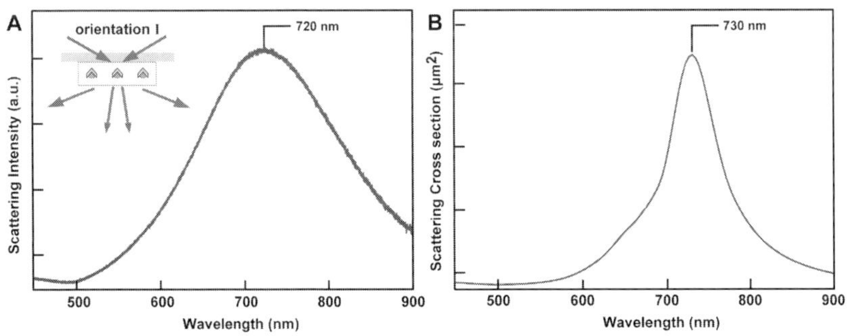

Figure 11.6 (A) Optical properties of arrays of 100-nm Au pyramids whose base plane is parallel to the substrate and whose tips point toward the optical axis of the microscope (orientation I). **(B)** Calculated scattering cross section of a Au pyramidal shell structure in this orientation and whose polarization vector was parallel to the base.

We first characterized a planar array of 100-nm Au pyramids ($t = 50$ nm) in PDMS whose tips were oriented perpendicular to the surface and whose base planes were oriented perpendicular to the optical axis of the microscope (orientation I). These pyramids supported only a single resonance around 720 nm (Fig. 11.6A). DDA calculations of a single pyramid indicated that the lone scattering peak originated from a dipole resonance (Fig. 11.6B), which suggested that the 100-nm pyramidal shells were not large enough to support multipolar resonances. This result is consistent with other reports that particles smaller than 100 nm typically do not usually exhibit multipolar plasmon resonances.[24]

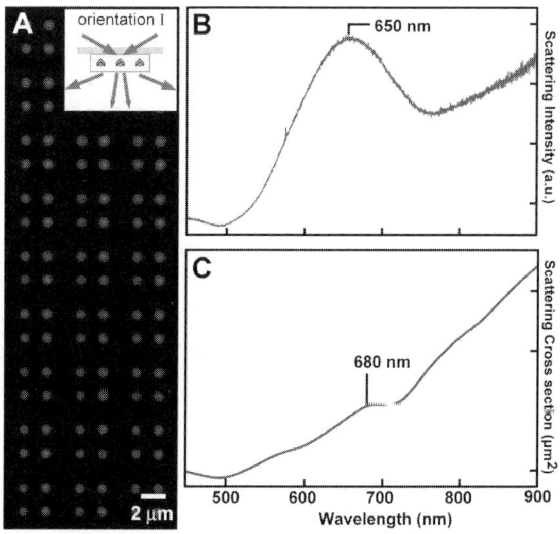

Figure 11.7 (A) DF microscope image of encapsulated 250-nm Au pyramids whose base plane is parallel to the substrate and whose tips point toward the optical axis of the microscope (inset, orientation I). **(B)** Scattering spectrum of the array of Au pyramids. **(C)** Calculated scattering cross section of a single Au pyramidal shell in this orientation with the polarization parallel to the base.

Planar arrays of 250-nm Au pyramids ($t = 50$ nm) in orientation I appeared as red spots in the DF micrograph (Fig. 11.7A). The spot sizes of the scattered light were uniform, and the color of every spot was the same. The scattered spectrum obtained with unpolarized white light exhibited a strong peak at red wavelengths (~ 650 nm) and another that appeared to extend into the NIR region (Fig. 11.7B). The optical response of this array did not change, however, when the incident light was polarized because the pyramidal particles in this orientation were symmetric with respect to the optical axis of the microscope. Also, the scattering from the same array with tips pointing *away* from the light source (the PDMS film was flipped over) was identical to the spectra in Fig. 11.7B.

Figure 11.7C shows the calculated spectrum of a single Au pyramidal shell, which exhibited a weak peak at 680 nm superimposed on a background that increased with increasing wavelength. An analysis of the induced polarizations indicated that the peak at 680 nm was from a quadrupole resonance—localized in the base plane of the pyramid—and that the background increased at longer wavelengths because of the tail of a dipolar excitation that peaks in the NIR. Note also that these calculations determined the Rayleigh scattering integral cross-section, which is not the same as the angle-resolved cross-section measured in the scattering spectra; however, the same resonance structure should be present. Figures 11.7B and 11.7C indicate that experiment and theory are in qualitative agreement, especially regarding the position of the quadrupole resonance, with the main difference being the intensity of the resonance. This difference could be attributed to several factors, including the difference between the calculated integral and angle-resolved cross sections noted above or the sensitivity of the calculations to subtle structural features of the pyramids.

11.9 Scattering Properties of Arrays of Pyramids with Tips Pointed Parallel to a Surface (Orientation II)

To investigate whether the optical properties of the pyramids depended on polarization, we exploited the flexible nature of PDMS and sliced the *h*-PDMS film containing the Au pyramids into thin (500 μm) sections. One of these cross-sections was placed on a glass substrate such that the planar array of pyramids was *parallel* to the optical axis of the microscope; the tips were oriented parallel to and the base planes were pointed perpendicular to the substrate (orientation II) (Fig. 11.8A, inset). Since the orientation of the particles is now *not* symmetric with respect to the incident light, we can study the effects of polarization on the scattering spectra. Figure 11.8A shows a DF image of a thin cross-section of an array with only the bottom layer of pyramids in the depth of focus (hence the scattered light is collected only from a single layer of pyramids). Excitation by unpolarized white light produced only broad features in the scattering spectra, unlike in orientation I, where unpolarized as well as polarized light produced a defined quadrupole peak.

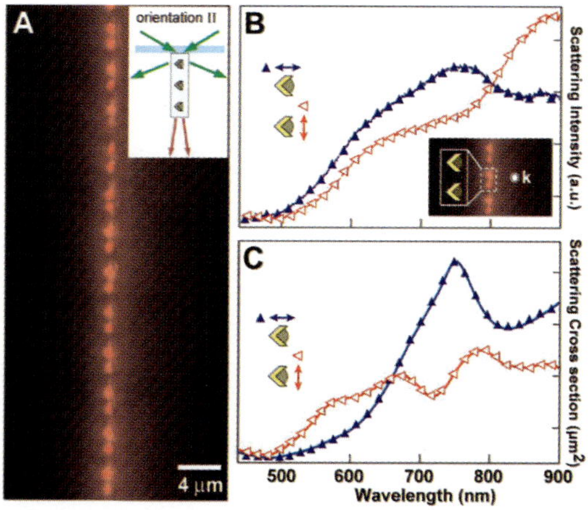

Figure 11.8 (A) DF microscope image of encapsulated 250-nm Au pyramids whose base plane is perpendicular to the substrate and whose tips are oriented perpendicular to the optical axis of the microscope (inset, orientation II). **(B)** Scattering spectra of the Au pyramids in **(A)** illuminated with white light that is polarized parallel to the pyramid base (▲) and polarized perpendicular to the pyramid base (◁). The inset shows how the pyramids in the array are oriented in **(A)**, and k denotes the propagation wavevector, which is parallel to and down the array plane. **(C)** Calculated scattering cross section of a single Au pyramidal shell in this orientation for two perpendicular polarizations.

Figure 11.8B shows the resulting spectra for the cases where the polarization vector is parallel (▲) and perpendicular (◁) to the pyramidal base; all polarization directions were defined relative to the base plane. Rotating from parallel to perpendicular polarization produced a color change in the scattered light from deep-red to light-red, which corresponded to the appearance of a resonance peak around 750 nm for the parallel case, and 880 nm for the perpendicular case. In good agreement with experiment, the calculated spectrum shows a strong resonance at 750 nm for parallel polarization (Fig. 11.8C). A polarization vector analysis indicates that this peak is the *same* quadrupole mode localized in the base plane as the one observed for orientation I (Fig. 11.7C). The calculated spectrum for the perpendicular polarization indicates the presence of several resonances with an overall envelope that is qualitatively similar to the measured result, except that the intensity of the peak at approximately 880 nm is too weak. Similar to the comparison of results for pyramids in orientation I, the resonance poles of the theoretical calculation matched experiment well, but there were some differences in overall peak intensity.

It is interesting that the spectrum for polarization parallel to the base in Fig. 11.8B is red-shifted from the spectrum in Fig. 11.7B — even though they appear to be optically equivalent. In both cases, the polarization vector is parallel to the base, but the wavevectors relative to the base plane are different: in Fig. 11.8B the wavevector is parallel, but in Fig. 11.7B the wavevector is perpendicular. *This wavevector-dependence only arises when particles are large compared to the wavelength of light*, because the electric field shows significant oscillations along the propagation direction. For example, when the wavevector is perpendicular to the base, the pyramids are relatively "thin," and the excitation field is in-phase at the base of the particle; however, when the wavevector is parallel to the base, the pyramids are relatively "thick," and the excitation field is no longer completely in-phase. These differences can lead to the excitation of different superpositions of multipoles, which has the net effect of shifting the nominal wavelength of the dominant multipole. For the 250-nm Au pyramids, the strongest resonance is the quadrupole mode. In particular, excitation parallel to the pyramid base should result in a stronger admixture of *dipolar* excitation, leading to a peak that is red-shifted (Fig. 11.8B) compared to excitation perpendicular to the base (Fig. 11.7B). This comparison of experiment and theory is probably the clearest example where this effect has been observed.

11.10 Conclusions

Gold nanopyramids are a unique class of metal nanostructures that exhibit orientation and wavevector-dependent, multi-polar LSP resonances. The scattering properties of individual and arrays of nanopyramids in different orientations matched well with DDA calculations. However, more work is needed to understand the plasmonic scattering modes of the nanopyramids in the NIR range. Because the relative orientation of the pyramids on a surface can be determined spectrally, these nanoparticles could be used for sensing and imaging applications by controlling the interaction between the faces of the pyramids and functionalized substrates. Also, arrays of pyramids could be used as multiplexed platforms for the detection of chemical and biological moieties or as surface-enhanced Raman substrates because of their sharp tips and edges.

References

1. Y. Xia and N. J. Halas, *Shape-Controlled Synthesis and Surface Plasmonic Properties of Metallic Nanostructures*, MRS Bull. 30, 338 (2005)
2. U. Jeong, X. Teng, Y. Wang, Y. Yang and Y. Xia, *Superparamagnetic colloids: controlled synthesis and niche applications*, Adv. Mater. 19, 33 (2007)
3. S. J. Oldenburg, G. D. Hale, J. B. Jackson and N. J. Halas, *Light scattering from dipole and quadrupole nanoshell antennas*, Appl. Phys. Lett. 75, 1063 (1999)
4. A. J. Haes, C. L. Haynes, A. D. McFarland, G. C. Schatz, R. P. van Duyne and S. Zou, *Plasmonic materials for surface-enhanced sensing and spectroscopy*, MRS Bull. 30, 368 (2005)
5. A. Yakimov and S. R. Forrest, *High photovoltage multiple-heterojunction organic solar cells incorporating interfacial metallic nanoclusters*, Appl. Phys. Lett. 80, 1667 (2002)
6. C. J. Murphy, T. K. Sau, A. M. Gole, C. J. Orendorff, J. Gao, L. Gou, S. E. Hunyadi and T. Li, *Anisotropic Metal Nanoparticles: Synthesis, Assembly, and Optical Applications*, J. Phys. Chem. B 2005, 13857 (2005)
7. L. J. Sherry, S.-H. Chang, G. C. Schatz, R. P. Van Duyne, B. J. Wiley and Y. Xia, *Localized Surface Plasmon Resonance Spectroscopy of Single Silver Nanocubes*, Nano Lett. 5, 2034 (2005)
8. A. Tao, P. Sinsermsuksakul and P. Yang, *Polydedral Silver Nanocrustals with Distinct Scattering Signatures*, Angew. Chem. Int. Ed. 45, 4597 (2006)
9. J. Chen, J. M. McLellan, A. Siekkinen, Y. Xiong, Z. Y. Li and Y. J. Xia, *Facile Synthesis of Gold-Silver Nanocages with Controllable Pores on the Surface*, J. Am. Chem. Soc. 128, 14776 (2006)
10. C. L. Nehl, N. K. Grady, G. P. Goodrich, F. Tam, N. J. Halas and J. H. Hafner, *Scattering Spectra of Single Gold Nanoshells*, Nano Lett. 4, 2355 (2004)

11. C. L. Nehl, H. Liao and J. H. Hafner, *Optical Properties of Star-Shaped Gold Nanoparticles*, Nano Lett. 6, 683 (2006)
12. R. Bukasov and J. S. Shumaker-Parry, *Highly-Tunable Infrared Extinction Properties of Gold Nanocrescents*, Nano Lett. 7, 1113 (2007)
13. B. K. Canfield, S. Kujala, M. Kauranen, K. Jefimovs, T. Vallius and J. Turunen, *Remarkable polarization sensitivity of gold nanoparticle arrays*, Appl. Phys. Lett. 86, 183109 (2005)
14. E. M. Hicks, S. Zou, G. C. Schatz, K. G. Spears, R. P. Van Duyne, L. Gunnarson, T. Rindzevicius, B. Kasemo and M. Kall, *Controlling Plasmon Line Shapes through Diffractive Coupling in Linear Arrays of Cylindrical Nanoparticles Fabricated by Electron Beam Lithography*, Nano Lett. 5, 1065 (2005)
15. J. S. Shumaker-Parry, H. Rochholz and M. Kreiter, *Fabrication of Crescent-Shaped Optical Antennas*, Adv. Mater. 17, 2131 (2005)
16. G. L. Liu, Y. Lu, J. Kim, J. C. Doll and L. P. Lee, *Magnetic Nanocrescents as Controllable Surface-Enhanced Raman Scattering Nanoprobe for Biomolecular Imaging*, Adv. Mater. 17, 2683 (2005)
17. K. A. Willets and R. P. Van Duyne, *Localized Surface Plasmon Resonance Spectroscopy and Sensing*, Ann. Rev. Phys. Chem. 58, 267 (2006)
18. S. R. Nicewarner-Pena, R. G. Freeman, B. D. Reiss, L. He, D. J. Pena, I. D. Walton, R. Cromer, C. D. Keating and M. J. Natan, *Submicrometer Metallic Barcodes*, Science 294, 137 (2001)
19. J. Henzie, E.-S. Kwak and T. W. Odom, *Mesoscale Metallic Pyramids with Nanoscale Tips*, Nano Lett. 5, 1199 (2005)
20. M.-C. Daniel and D. Astruc, *Gold Nanoparticles: Assembly, Supramolecular Chemistry, Quantum-Size-Related Properties, and Applications toward Biology, Catalysis, and Nanotechnology*, Chem. Rev. 104, 293 (2004)
21. T. Jensen, M. Duval, K. Kelly, A. Lazarides, G. Schatz and R. Van Duyne, *Nanosphere Lithography: Effect of the External Dielectric Medium on the Surface Plasmon Resonance Spectrum of a Periodic Array of Silver Nanoparticles*, J. Phys. Chem. B 103, 9846 (1999)
22. C. F. Bohren and D. R. Huffman, *Absorption and Scattering of Light by Small Particles* (Wiley, New York, 1998)
23. J. Henzie, K. L. Shuford, E.-S. Kwak, G. C. Schatz and T. W. Odom, *Manipulating the Optical Properties of Pyramidal Nanoparticle Arrays*, J. Phys. Chem. B 110, 14028 (2006)

24. P. K. Jain, K. S. Lee, I. H. El-Sayed and M. A. El-Sayed, *Calculated Absorption and Scattering Properties of Gold Nanoparticles of Different Size, Shape, and Composition: Applications in Biological Imaging and Biomedicine*, J. Phys. Chem. B 110, 7238 (2006)
25. A. S. Kumbhar, M. K. Kinnan and G. Chumanov, *Multipole Plasmon Resonances of Submicron Silver Particles*, J. Am. Chem. Soc. 127, 12444 (2005)
26. J. E. Millstone, S. Park, K. L. Shuford, L. Qin, G. C. Schatz and C. A. Mirkin, *Observation of a Quadrupole Plasmon Mode for a Colloidal Solution of Gold Nanoprisms*, J. Am. Chem. Soc. 127, 5312 (2005)
27. K. L. Shuford, M. A. Ratner and G. C. Schatz, *Multipolar excitation in triangular nanoprisms*, J. Chem. Phys. 123, 114713 (2005)
28. K. L. Kelly, E. Coronado, L. L. Zhao and G. C. Schatz, J. Phys. Chem. B 107, 668 (2003)
29. J. R. Krenn, G. Schider, W. Rechberger, B. Lamprecht, A. Leitner, F. R. Aussenegg and J. C. Weeber, *Design of multipolar plasmon excitations in silver nanoparticles*, Appl. Phys. Lett. 77, 3379 (2000)
30. S. Malynych and G. Chumanov, *Light-Induced Coherent Interactions between Silver Nanoparticles in Two-Dimensional Arrays*, J. Am. Chem. Soc. 125, 2896 (2003)
31. J. Henzie, J. E. Barton, C. L. Stender and T. W. Odom, *Large-Area Nanoscale Patterning: Chemistry Meets Fabrication*, Acc. Chem. Res. 39, 249 (2006)
32. W. R. Childs and R. G. Nuzzo, *Large-Area Patterning of Coinage-Metal Thin Films Using Decal Transfer Lithography*, Langmuir 21, 195 (2005)
33. T. W. Odom, J. C. Love, D. B. Wolfe, K. E. Paul and G. M. Whitesides, *Improved Pattern Transfer in Soft Lithography Using Composite Stamps*, Langmuir 18, 5314 (2002)
34. W. Hasan, J. Lee, J. Henzie and T. W. Odom, *Selective Functionalization and Spectral Identification of Au Nanopyramids*, submitted to J. Phys. Chem. C, 111, 1716 (2007)

12

PROPERTIES OF GOLD NANOANTENNAS IN THE INFRARED

F. Neubrech, J. Aizpurua, S. Karim, T. W. Cornelius and A. Pucci

With infrared spectroscopic microscopy using synchrotron light, resonant light scattering from single gold nanowires was investigated. The nanowires with diameters in the range of 100 nm were prepared by electrochemical deposition in polycarbonate etched ion-track membranes and transferred onto infrared-transparent substrates. For a few microns long nanowires, antenna-like plasmon resonances were observed in agreement with light scattering calculations. The resonances are dependent on wire length, wire shape, and on the dielectric surroundings of the nanowire. Electromagnetic far-field enhancement at resonance points to the ability of the nanowire to confine light on the nanoscale. This effect can be exploited for surface enhanced infrared absorption.

12.1 Introduction

Below the onset of interband transitions, the infrared (IR) optical properties of metals are determined by the collective oscillations of free charge carriers called plasmons. The Drude dielectric function

$$\varepsilon(\omega) = \varepsilon_\infty - \frac{\omega_p^2}{\omega(\omega + i\omega_\tau)} \tag{1}$$

is a good description of that behaviour.[1-5] Effects from interband transitions on the background polarizability are included in ε_∞. The Drude parameters ω_p and ω_τ describe the plasma frequency and the relaxation rate of the free charge carriers, respectively. With ω_p much higher than IR frequencies and $\omega_\tau << \omega_p$ the Drude dielectric function has a strong negative real part in the IR and corresponds to a considerably high polarizability. Above (circular) frequency $\omega = \omega_p / \sqrt{\varepsilon_\infty}$ (the so-called plasma edge) the high metallic reflectivity drops down to values that depend on background polarizability and interband transitions. The high reflectivity at lower frequencies is related to the small penetration depth (skin depth) of the IR light into the metal. For ω_τ in the mid IR (typical for iron or defect-rich noble metals) the skin depth in the IR[6] is

$$\delta = \frac{c}{\omega_p}\sqrt{\frac{2\omega_\tau}{\omega}} \tag{2}$$

with c as the velocity of light. Typical values for skin depths in the mid IR are 10 to 50 nm for metals. Therefore, metal films of a few nm in thickness are partially transparent in the IR. This behaviour extends to the visible range where (for sufficiently large ω_p) in case of oblique incidence of p-polarized light the transmittance spectra of ultrathin metal films may show the additional plasmon resonance around $\omega = \omega_p / \sqrt{\varepsilon_\infty}$ due to the dispersion of surface plasmon polaritons.[7] This resonance corresponds to collective charge carrier oscillations (plasmons) in perpendicular direction for which the film boundaries are effective. For metal particles the size and shape determine the boundary conditions for the plasmon oscillations. Vice versa, the optical resonances can be tuned via shape and size. The tunable optical properties of metal nanoparticles and their applications have been a topic of dramatically increasing interest over the last years[8-11]. In particular, this interest is related to the considerable electromagnetic field enhancement in the near-field of a metal nanoparticle at resonance. As an example, it can be used to control field amplification in surface enhanced Raman scattering (SERS)[12,13]. Extremely high near-field enhancement was calculated for the region of a slit between nanorods[14]. Related to this, electromagnetic resonances of nano-objects can be exploited to confine light to a volume of sub-wavelength dimensions[15,16]. Arrays of nanosized objects exhibit interesting photonic properties[17] and composites of nanostructures are the focus of search for metamaterials[18-20]. Compared to antenna-like resonances in the visible range[21], electromagnetic field enhancement by metal antennas is higher in the infrared because of the strong negative part of the metal dielectric function and the high aspect ratios that can be produced. The resonant modes of collective charge carrier motion that can be excited by electromagnetic radiation with wavelength much larger than the particle size are mainly determined by the material properties of the nanoparticles.[22] Going from spherical particles to prolate spheroids, the shape dependence of the resonances becomes much stronger[23,24] mainly for electric field along the longer particles axis. The plasmon resonance splits into branches with a low-energy mode, which can be detected in the mid-infrared for an adequate length of the nanoparticle[25]. The larger the aspect ratio (long axis L / small axis D) of the particle the lower the frequency of this mode. When the wire length approaches the range of the exciting wavelength, the effects of retardation dominate the resonance condition and the limited speed of light leads to a direct dependence of the resonance frequency on the absolute size of the particle. This is well known from purely classical theory of scattering of electromagnetic waves by ideal metal objects[26]. Here antenna resonances occur if the length L of an infinitely thin wire roughly matches with multiples of the wavelength λ, *i.e.*

$$L = \frac{\lambda}{2n} \cdot m \,, \tag{3}$$

where m is a natural number and n is the refractive index of the surrounding medium. The fundamental mode of the resonance corresponds to $m = 1$, higher dipole-like excitations correspond to odd m. If the diameter D of the antennas is no longer negligible compared to the wavelength, the resonance frequency depends also on D[26]. Further, the skin effect has to be considered if D is in the range of the skin depth δ. In that case, the skin effect has an impact on the dependence of the antenna resonance on the material properties[27] and introduces a qualitative change of the resonance relation[14]. The change caused by the influence of δ on the resonance condition is opposite to the geometrical one caused by a non-negligible D. Through the ratio δ/D the Drude parameters influence the resonance curve of a nanoantenna. In particular, high defect density inside the metal particles would increase ω_τ and thus the ratio δ/D. In addition to the above mentioned effects, via the boundaries, the polarizability of the surrounding medium also affects the optical properties of metal nanoparticels[28]. In order to take into account the effect of a substrate (with the refractive index ε_s) an effective medium completely embedding the wires is a good approximation. With its effective dielectric constant

$$\varepsilon_{eff} = \frac{1}{2}(1+\varepsilon_s) \tag{4}$$

and $n = n_{eff} = \sqrt{\varepsilon_{eff}}$ the influence of the substrate on optical spectra can be described well[22,29].

12.2 Measurement of Antenna Resonances in the Infrared

The clearest experimental result is expected for individual gold wires with a cylindrical shape, a low defect density, and a clean surface. Gold is preferred because of its stability under ambient conditions. Gold wires with diameters between 60 and 200 nm were prepared, for example, by electrochemical deposition in etched ion-track membranes[30,31]. Fig. 12.1 shows a scanning electron microscopy (SEM) image of a gold nanowire (with D = 100 nm), displaying its smooth contour and regular shape. Transmission electron microscopy studies proved that gold nanowires deposited under the conditions mentioned above are polycrystalline[30]. After formation of wires with a controlled length, the polymer membrane was dissolved in dichloromethane in order to achieve clean nanowires.

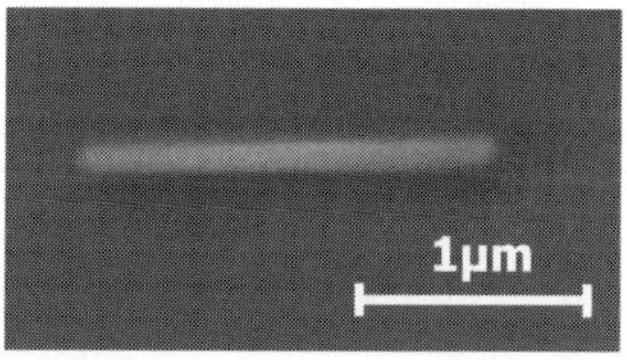

Fig. 12.1 SEM image of a gold wire with 100 nm in diameter, see text.

For IR spectroscopy these wires were transferred to different IR transparent substrates (ZnS, KBr, and CaF_2), with different refractive indices n_s (n_{ZnS} = 4.84 at 10 µm, n_{KBr} = 1.53, and n_{CaF2} = 1.4 at 5 µm). During the sample preparation process, the several microns long wires broke into pieces with lengths between several hundred nm and few µm, and with tip ends determined by the grain-boundary structure of the wires.

IR spectroscopic microscopy of single gold nanowires, produced as explained above, was performed at the IR beamline of the synchrotron light source ANKA (Angströmquelle Karlsruhe) at the Forschungszentrum Karlsruhe. In contrast to wire-array studies, the individual wires were randomly placed on the IR transparent substrate and, therefore, their position needed to be located. At the beamline this and the length determination were done with optical microscopy with light (which gives an error of up to 8%). Then the IR-beam spot with diameter of about 8 µm was centered on the selected nanowire and IR-transmittance spectra (sample spectra) were recorded. Subsequently, in order to eliminate environmental effects like beam profile and substrate inhomogeneities, reference spectra were taken at least 10 µm away from any nanowire. The spectroscopic measurements were done with a Fourier-transform infrared (FT-IR) spectrometer (Bruker IFS 66 v/S) and a LN2-cooled mercury-cadmium telluride (MCT) detector, which collects light normally transmitted through the sample area. A small fraction of the light scattered away from the normal direction was also detected due to the collection lens (Schwarzschild objective, numerical aperture: 0.52), but its intensity was negligibly small. For both sample and reference measurements IR spectra were recorded by acquisition of at least 10 scans in the spectral range from 600 to 7000 cm^{-1} with a resolution of at least 16 cm-1. For further analysis of the transmitted light, a polarizer was inserted in the optical path before the sample.

For nanowires with a length of a few microns significant fundamental antenna-like plasmon resonances appear in the relative transmittance spectra (ratio of the sample and reference spectra, see Fig. 12.2). Apart from the fundamental resonance, much weaker structures of resonant extinction can be found at higher wavenumbers. Usually they are attributed to higher order resonances. The fundamental resonance was observed only for electrical field parallel to the long axis of the wire (parallel polarization). For polarization perpendicular to the long wire axis (perpendicular polarization), the IR signal was below the noise level. That behaviour proves the dipole character of the excitation attributed to $m = 1$.

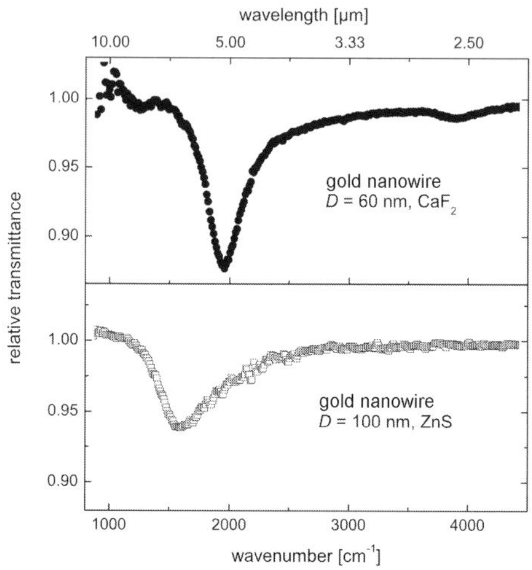

Fig. 12.2 IR transmittance spectra of two individual gold nanowires on different substrates (as indicated) normalized to the transmittance of the bare substrate. At the lower wave numbers the CaF_2 substrate arrives at the multi-phonon edge, which lowers the photon intensity at the detector, which gives rise to noise in the spectrum. The electric field was parallel to the wire.

From the measured relative transmittance spectra the ratio of the extinction cross section σ_{ext} of a single gold nanowire to its geometric cross section σ_{geo} can be estimated. As we already reported in reference 29, we use the relation

$$\frac{\sigma_{ext}}{\sigma_{geo}} = A_0(1 - T_{rel}) \cdot (n_s + 1) \cdot \frac{1}{(2LD)}, \tag{5}$$

where the substrate screening is taken into account in analogy to the normal transmittance change by a thin film on a substrate[32] in comparison to a free-standing film[33]. Inserting A_0 as the spot size, $(1\text{-}T_{rel})$ the transmittance change in the spectrum, n_s for the relevant substrate, L from dark-field microscopy in the visible spectral range, and D (known from wire-growth process) lead to $\sigma_{ext} / \sigma_{geo}$ as shown for one example in Fig. 12.3. Any value $\sigma_{ext} / \sigma_{geo} > 1$ means an extinction of intensity above pure shadowing, which indicates a local field enhancement in the vicinity of the nanowire. From the cross section ratio $\sigma_{ext} / \sigma_{geo}$ the spatially averaged field enhancement in the far-field ($\sqrt{\sigma_{ext} / \sigma_{geo}}$) can be calculated.

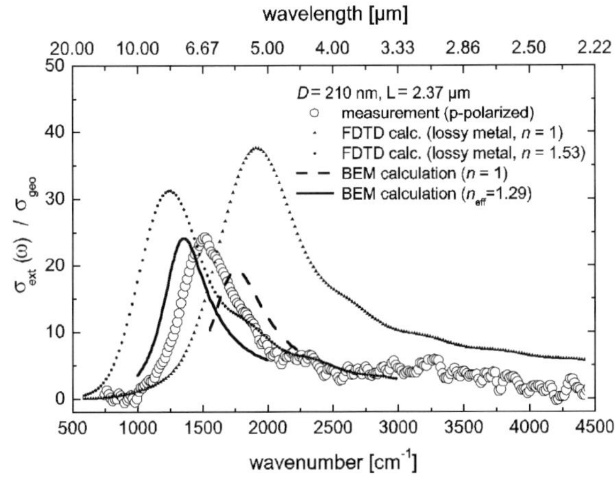

Fig. 12.3 Extinction cross section (normalized to geometric cross section): experimental curve for a gold nanowire on KBr compared to calculations for different effective refractive indices around the nanowire, see text.

The experimental spectrum that is shown in Fig. 12.3 represents one of our measurements and reveals far-field enhancement by a factor of about 5 at the resonance maximum. The other curves were derived with two different theoretical methods. In both the methods the gold nanowire is modelled as a rod with hemispherical tip ends completely embedded in a medium (background). The far-field observation point in both calculations is in the plane defined by the rod axis and the light propagation axis. Those resonance curves marked by "FDTD calc." were modelled with the software CST Microwave studio[34] for antenna problems and were based on a finite difference time domain (FDTD) algorithm[35]. In this computational approach, Maxwell's time dependent differential equations are solved over a three-dimensional grid. In our calculations one single gold nanowire was illuminated by a plane wave polarized along the long wire axis, and the antenna radiated into free space. The curves marked with "BEM calculations" were calculated with the exact boundary element method (BEM) including retardation[14,36]. Maxwell's equations for inhomogeneous media with sharp boundaries were solved in terms of charges and currents distributed on the surfaces and interfaces. Boundary conditions are imposed via surface integrals along the boundaries between different media. Each region is characterized by a local dielectric function. The external fields interact self-consistently with the induced boundary charges and currents, which are determined by discretizing the surface integrals and solving the appropriate matrix equations. In this approach, the scattered field due to an incident external field is calculated directly.

In the FDTD calculations, we described the metal of the gold nanowires as a lossy one. The "lossy metal" model neglects the imaginary part of the conductivity. Changing the background material from air to a dielectric one (KBr) a shift of the resonance maximum towards smaller wavenumbers is observed (dotted line). Obviously, neither the calculation based on vacuum as background nor the curve with background KBr describes the measured data. The resonance curves calculated with the boundary element method (BEM) are in better agreement with the experiment. Using an effective refractive index $n_{\mathrm{eff}} =$ 1.29 the best agreement, in strength and shape, of the resonance curve with the experimental data is obtained. Only the position of the extinction maximum differs slightly, which can be related to the error in the real length of the wire.

Looking at Fig. 12.4, where we display the resonance wavelengths, from the measured spectra for nanowires with comparable diameters on KBr, systematic deviation from simple antenna theory (black line) is obvious. When using the value of the refractive index of KBr in Eq .3 the calculated curve (dotted line) seems to fit our measured data. But this in conflict with the fact that the wires are not embedded in this material. On the other hand, BEM calculations for wires embedded in an effective medium with refractive index $n_{\mathrm{eff}} = \sqrt{\varepsilon_{\mathrm{eff}}}$ yield good agreement with the measured data. Also the effect of a cover medium is well described by the respective $\varepsilon_{\mathrm{eff}} = (\varepsilon_{\mathrm{m}} + \varepsilon_{\mathrm{s}})/2$ (for surrounding medium with dielectric constant ε_{m}) as it is shown by the coverage of individual gold nanowires on different substrates (KBr, ZnS) with paraffin wax. Due to the paraffin coverage the resonance is shifted to lower frequencies[29].

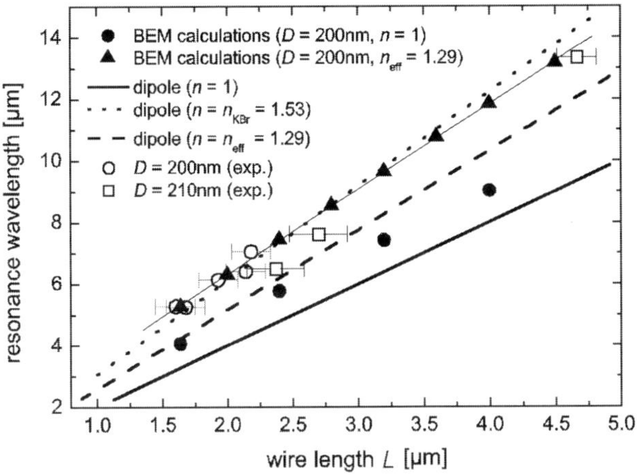

Fig. 12.4 Resonance wavelength versus wire length L for gold nanowires with comparable diameter on KBr. The figure shows experimental data (open symbols), ideal-antenna model calculations for wires embedded in different surroundings (lines), and detailed BEM calculations. The lines corresponding to Eq. 3 (m=1) are shown for comparison.

The theoretical ratio is $n_{\mathrm{eff}}^{\mathrm{air}} / n_{\mathrm{eff}}^{\mathrm{paraffin}} = 0.875$ on KBr and 0.923 on ZnS (with $\varepsilon_{\mathrm{m}} = 2.02$ for paraffin), respectively. These ratios are in reasonable accord with the results obtained from the spectral shift of the resonance frequencies before and after evaporating paraffin.

12.3 Surface Enhanced Infrared Absorption (SEIRA)

Field enhancement in the IR can be exploited for surface enhanced infrared absorption (SEIRA)[37,38] of molecules. In case of sufficiently strong near-field enhancement, SEIRA would allow the detection of vibrational fingerprints of single molecules on nanoantennas, for example, by studying light scattering with aperture-less scanning near-field microscopy[39].

Until now SEIRA is less well-known than SERS since its signal enhancement was measured to be only three orders of magnitudes[37,38]. This is much less than the SERS enhancement which is many more orders of magnitude. This difference is related to the different influence of local electromagnetic field enhancement. In SEIRA the squared near-field enhancement is relevant, whereas in SERS, which is a scattering process, two kinds of squared field enhancements are multiplied (one for the incident field and the other one for the scattered field).[40] Usually, the electromagnetic field enhancement is obtained with metal nanoparticles or nanostructured metal films. Their plasmonic resonances give rise to near-field enhancement as the main effect in SEIRA and SERS[41]. Other contributions to the observed enhancement of vibration signals are less important (concerning the contribution to enhancement) and less well understood. They are related to chemical interactions and non-adiabatic effects and are relevant for the first adsorbate layer only.[42]

The near-field enhancement is strongest in the proximity of sharp edges or tips. For SERS this tip effect was already studied by several groups.[12,43] For nanoantennas in the IR such tip effect should be much more like a lightning rod effect. This would means huge near-field enhancement at the tip ends of an antenna at resonance, much larger than three orders of magnitude as measured as the spatially averaged enhancement for adsorbates on metal-island films compared to the same adsorbate layer without metal islands.

Acknowledgements

F. N. and A. P. gratefully acknowledge technical support by M. Süpfle, helpful discussions with A. Otto, B. Gasharova, and Y.-L. Mathis. J.A. acknowledges software access from F.J. García de Abajo, discussions with G.W. Bryant, and financial support from Gipuzkoa Foru Aldundia (Gipuzkoa Fellows program) and the Government of the Basque Country (NANOTRON project).

References

1. G. Fahsold, A. Bartel, O. Krauth, N. Magg and A. Pucci, IR-optical properties of ultrathin Fe films on MgO(001) beyond the percolation threshold, Phys. Rev B 61, 14108 (2000)
2. G. Fahsold, M. Sinther, A. Priebe, S. Diez and A. Pucci, Adsorbate-induced changes in the broadband infrared transmission of ultrathin metal films, Phys. Rev. B 65, 235408 (2002)
3. G. Fahsold and A. Pucci, Non-contact measurement of thin-film conductivity by ir spectroscopy, Advances in Solid State Physics (ed. by B. Kramer, Springer press) 43, 833 (2003)

4. A. Pucci, F. Kost, G. Fahsold and M. Jalochowski, Infrared spectroscopy of Pb layers growth on Si (111), Phys. Rev B 74, 125428 (2006)
5. F. Meng and A. Pucci, Growth of silver on MgO (001) and infrared optical properties, phys. stat. sol. (a) in press (2007)
6. F. Abeles, Optical properties of metals, in Optical properties of solids, ed. by F. Abeles (North-Holland Publishing, Amsterdam, 1972), pp. 93
7. Y. Borensztein, M. Roy and R. Alameh, Threshold and Linear Dispersion of the plasma resonance in Thin Ag Films, Europhysics Lett. 31, 311 (1995)
8. S. Link and M. A. El-Sayed M A 1999, Spectral properties and relaxation dynamics of surface plasmon electronic oscillations in gold and silver nano-dots and nano-rods, J. Phys. Chem. B 103, 8410 (1999)
9. N. Félidj, J. Aubard, G. Lévi, J. R. Krenn, M. Salerno, G. Schider, B. Lambrecht, A. Leitner and F.R. Aussenegg, Controlling the optical response of regular arrays of gold particles for surface-enhanced Raman scattering, Phys. Rev. B 65, 075419 (2002)
10. G. Schider, J. R. Krenn, A. Hohenau, H. Ditlbacher, A. Leitner, F. R. Aussenegg, W. L. Schaich, I. Puscasu, B. Monacelli and G. Boreman, Plasmon dispersion relation of Au and Ag nanowires, Phys. Rev. B 68, 155427 (2003)
11. C. Soennichsen, T. Franzl, T. Wilk, G. von Plessen, J. Feldmann, O. Wilson and P. Mulvaney, Drastic reduction of plasmon damping in gold nanorods, Phys. Rev. Lett 88, 077402 (2002)
12. J. Grand, M. Lamy de la Chapelle, J. L. Bijeon, P. M. Adam, A. Vial and P. Royer, Role of localized surface plasmons in surface-enhanced Raman scattering of shape-controlled metallic particles in regular arrays, Phys. Rev. B 72, 033407 (2005)
13. H. Xu, E. J. Bjerneld, M. Kaell and L. Boerjesson, Spectroscopy of single hemoglobin molecules by surface enhanced Raman scattering, Phys. Rev. Lett. 83 4357 (1999)
14. J. Aizpurua, G. W. Bryant, L. J. Richter, F. J. García de Abajo, B. K. Kelley and T. Mallouk, Optical properties of coupled metallic nanorods for field-enhanced spectroscopy, Phys. Rev. B 71, 235420 (2005)
15. E. Cubukcu, E. A. Kort , K. B. Crozier and F. Capasso, Plasmonic laser antenna, Appl. Phys. Lett. 89, 093120 (2006)

16. A. Maier, P. G. Kik and H. A. Atwater, Observation of coupled plasmon-polariton modes in Au nanoparticle chain waveguides of different lengths: Estimation of waveguide loss, Appl. Phys. Lett. 81, 1714 (2002)
17. M. Naruse, T. Kawazoe, S. Sangu, K. Kobayashi, and M. Ohtsu, Optical interconnects based on optical far- and near-field interactions for high-density data broadcasting, Opt. Express 14, 306 (2006)
18. R. Moussa, S. Foteinopoulou, L. Zhang, G. Tuttle, K. Guven, E. Ozbay and C. M. Soukoulis, Negative refraction and superlens behavior in a two-dimensional photonic crystal, Phys. Rev. B 71, 085106 (2005)
19. J. B. Pendry, Negative Refraction Makes a Perfect Lens, Phys. Rev. Lett. 85, 3966 (2000)
20. V. A. Podolskiy, A. K. Sarychev, E. E. Narimanov and V. M. Shalaev, Resonant light interaction with plasmonic nanowire systems, J. Opt. A: Pure Appl. Opt. 7, 32 (2005)
21. J. P. Schuck, D. P. Fromm, A. Sundaramurthy, G. S. Kino and W. E. Moerner, Improving the Mismatch between Light and Nanoscale Objects with Gold Bowtie Nanoantennas, Phys. Rev. Lett. 94, 017402 (2005)
22. for example: M. M. Dvoynenko, A. V. Goncharenko, V. R. Romaniuk V R and E. F. Venger,) Effects of dimension on optical transmittance of semicontinuous gold, Physica B 299 88 (2001)
23. J. Grand, P. M. Adam, A. S. Grimault, A. Vial, M. Lamy de la Chapelle, J. L. Bijeon, S. Kostcheev and P. Royer, Optical Extinction Spectroscopy of Oblate, Prolate, and Ellipsoid Shaped Gold Nanoparticles: Experiments and Theory, Plasmonics 1, 135 (2006)
24. A. Bouhelier, R. Bachelot, G. Lerondel, S. Kostcheev, P. Royer and G. P. Wiederrecht, Surface Plasmon Characteristics of Tunable Photoluminescence in Single Gold Nanorod, Phys Rev. Lett 95, 267405 (2005)
25. K. B. Crozier, A. Sundaramurthy, G. S. Kino and C. F. Quate, Optical antennas: Resonators for local field enhancement, J. Appl. Phys. 9, 4632 (2003)
26. T. G. Ruck, Radar Cross Section Handbook (Plenum Press, New York, 1970)
27. R. B. Dingle, The Electrical Conductivity of Thin Wires, Proceedings of the Royal Society of London, Series A, Mathematical and Physical Sciences 201 1067 (1950)

28. J. Mock, D. R. Smith and S. Schultz, Local refractive index dependence of plasmon resonance spectra from individual nanoparticles, Nanoletters 3, 485 (2003)
29. F. Neubrech, T. Kolb, R. Lovrincic, G. Fahsold, J. Aizpura, M. E. Toimil-Molares, S. Karim, T. W. Cornelius, R. Neumann and A. Pucci, Resonances of individual metal nanowires in the infrared, Appl. Phys. Lett. 89, 253104 (2006)
30. J. Liu, J. L. Duan, M. E. Toimil-Molares, S. Karim, T. W. Cornelius, D. Dobrev, H. J. Yao, Y. M. Sun, M. D. Hou, D. Mo, Z. G. Wang and R. Neumann, Electrochemical fabrication of single crystalline and polycrystalline Au nanowires: the influence of deposition parameters, Nanotechnology 17, 1922 (2006)
31. S. Karim, M. E. Toimil-Molares, F. Maurer, G. Miehe, W. Ensinger W, J. Liu, T. W. Cornelius and R. Neumann, Synthesis of gold nanowires with controlled crystallographic characteristics, Appl. Phys. A 84, 403 (2006)
32. U. Teschner and K. Huebner, IR-spectroscopic data of thin insulating films on semiconductors. New methods of interpretation and analysis. phys. stat. sol. (b) 159, 917 (1990)
33. D. W. Berreman, Infrared Absorption at Longitudinal Optic Frequency in Cubic Crystal, Phys. Rev. 130 2193 (1963)
34. CST Computer Simulation Technology, Darmstadt, Germany
35. K. S. Yee, Numerical solution of boundary value problems involving Maxwell's equations in isotropic media, IEEE Trans. Antennas Propag. 14, 302 (1966)
36. F. J. García de Abajo and A. Howie, Relativistic Electron Energy Loss and Electron-Induced Photon Emission in Inhomogeneous Dielectrics, Phys. Rev. Lett. 80, 5180 (1998) and Retarded field calculation of electron energy loss in inhomogeneous dielectrics, Phys. Rev. B 75, 115418 (2002)
37. R. F. Aroca, D. J. Ross and C. Domingo, Surface-Enhanced Infrared Spectroscopy, Applied Spectroscopy 58, 324A (2004)
38. D. Enders and A. Pucci, Surface enhanced infrared absorption of octadecanethiol on wet chemically prepared Au nanoparticle films, Appl. Phys. Lett. 88, 184104 (2006)
39. A. Cvitkovic, N. Ocelic, J. Aizpurua, R. Guckenberger and R. Hillenbrand, Infrared Imaging of Single Nanoparticles via Strong Field Enhancement in a Scanning Nanogap, Phys. Rev. Lett. 97, 060801 (2006)

40. A.W. Wokaun in: H. Ehrenreich, F. Seitz, D. Turnbull (Eds.), Surface enhanced Electromagnetic Processes, Solid State Phys, vol. 38, (Academic Press, New York, 1984), p 223
41. A. Otto, I. Mrozek, H. Grabhorn, and W. Akemann, Surface enhanced Raman scattering, J. Phys.: Condens. Matter 4, 1143 (1992)
42. A. Otto, The chemical (electronic) contribution to surface-enhanced Raman scattering, J. Raman spectroscopy 36, 497 (2005)
43. N. Calander and M. Willander, Theory of surface-plasmon resonance optical field enhancement at prolate spheroids, J. Appl. Phys. 92, 4878 (2002)

13

THREE-DIMENSIONAL HOLOGRAPHIC POLYMERIC PHOTONIC CRYSTAL OPERATING IN THE OPTICAL COMMUNICATION WINDOW

Jiaqi Chen and Ray T. Chen

A polygonal prism-based holographic fabrication method has been demonstrated for a six layer three-dimensional face-centered-cubic (FCC)-type submicron polymeric photonic crystal using both positive and negative photoresists. Special fabrication treatments have been introduced to insure the survivability of the fabricated large area (cm^2) nano-structures. The SEM and diffraction results proved the good uniformity of the fabricated structures. With the proper design of the refraction prism we have achieved the required bandgap for S+C band (1460-1565nm) in the [111] direction. The simulation of the band structure is in good agreement with transmission and reflection spectra obtained by Fourier transform infrared spectroscopy (FTIR). The superprism effects around 1550nm wavelength for the fabricated 3D polymer photonic crystals have also been theoretically calculated and such effects can be used for beam steering purpose.

13.1 Introduction

The prediction and the confirmation that artificial periodic dielectric structures can be used to manipulate electromagnetic wave propagation have significantly affected the development of micro- and nano-optoelectronics[1-3]. In the literature there is a great diversity in the fabrication approaches to making two-dimensional and three-dimensional photonic crystals. Therefore, there is a need to improve the quality of such structures in terms of being more feature-size flexible, materials flexible and less time consuming. Despite the remarkable progress in the fabrication of two-dimensional photonic crystals, there remain significant challenges for the fabrication of 3D photonic crystals, especially for producing sub-micron periodicity for near-IR applications. Many 3D fabrication approaches have been studied on a number of different material platforms. Among them, layer-by-layer stacking using soft lithography or nanoimprint lithography, microassembly of planar semiconductor layers and multi-photon absorption have been investigated to create certain microstructures. However, for the aforementioned approaches, multiple processing steps are generally required and these are not suitable for larger scale (cm^2 size) fabrication.

Recently, after Campbell and coworkers[4] successfully demonstrated the holographic fabrication of 3D photonic crystals, holographic lithography has gained more attention in the fabrication of 3D polymeric photonic crystals. By controlling the propagation directions, phases and polarizations of multiple laser beams one can generate a desired 3D laser interference pattern to expose a photo-sensitive material such as photoresist and render the exposed areas soluble or insoluble depending on the material type in use. This process allows flexible design as well as control of the unit cell geometry and, therefore, of the desired optical properties. It has been proven theoretically that all fourteen Bravais lattices can be constructed using such a method. In the early days of this approach, multiple beams were formed by independent optical components. Such setups were complicated and required careful alignment. Recently, Divliansky et al.[5] introduced the use of a diffraction mask to create the four desired beams and Wu et al.[6] demonstrated another idea using a single refracting prism to split and combine the desired beams. With the use of a single optical element to split and recombine the laser beams, sophisticated anti-vibration system and alignment are avoided. However, the photonic bandgap and superprism effect of most of these devices did not fall in the optical communication windows due to the relatively large periodicities involved. In the experiments described here, based on the diffraction prism approach, we used a He-Cd laser to fabricate submicron 3D photonic crystal structures. We utilized a negative i-line resist (SU8) with the 325 nm wavelength for superprism application near 1550 nm.

13.2 Fabrication Setup and Device Structure Simulation Results

The optical setup for the fabrication using SU8 resist is illustrated in Figure 13.1(a). The UV (325nm) laser beam coming from the He-Cd laser was expanded using a UV objective lens. The spatial filter afterward cleaned up the unwanted scattering in the optical path to ensure a good beam quality. The laser beam was collimated by a large lens so as to have a diameter of 10 cm and it was incident at normal incidence on the specially designed top-cut prism. The prism was made from JGS1 material (with a refractive index of 1.482 at 320nm) to achieve a larger than 95% UV transmission and the surfaces of the top-cut prism had been polished to make the transmission intensity as large as possible. The three sidewalls of the top-cut prism created three side beams by beam refraction. The center beam of this umbrella interference configuration was formed from the central part of the collimated laser beam going directly through the top and bottom surfaces of the prism. The four beams created by the prism overlapped at the bottom of the prism. By controlling the height of the prism one could control the depth of overlap to ensure the photoresist was within the overlap volume. In our design shown in Figure 13.1(b) L1 = 1 cm, L2 = 4 cm and H = 1.5 cm, therefore, we achieved a cutting angle of 60 ° for the prism. Then the resulting wave vectors were $G_0 = \frac{2\pi}{\lambda}(0,0,-1)$, $G_1 = \frac{2\pi}{\lambda}(\text{-}0.3314,\text{-}0.1914,\text{-}0.9239)$, $G_2 = \frac{2\pi}{\lambda}(0.3314,\text{-}0.1914,\text{-}0.9239)$, $G_3 = \frac{2\pi}{\lambda}(0,0.3827,\text{-}0.9239)$, where λ= 325 nm for the use of SU8 resist and 442 nm for the use of AZ 4620 resist. The interference pattern can be calculated by

$$I(\vec{r}) = \sum_{i=0}^{N-1} E_i^2 + 2\sum_{i=1}^{N-1} E_0 \times E_i \times \vec{e}_0 \bullet \vec{e}_i \times \cos(\vec{K}_i \bullet \vec{r} + \theta_i)$$

$$+2\sum_{i>j=1}^{N-1} E_i \times E_j \times \vec{e}_i \bullet \vec{e}_j \times \cos(\vec{K}_{ij} \bullet \vec{r} + \theta_{ij}) \tag{13.1}$$

where $K_i = G_i - G_0$, $K_{ij} = K_i - K_j$ and θs are the phase differences of the beams.

For the negative photoresist a schematic drawing of the position of the prism, the glass substrate and the absorption plate is shown in figure 13.1(c). The bottom black glass plate was used to reduce the back-reflected light and all the interfaces used index matching liquid to avoid the unwanted reflections at the glass/air and polymer/air interfaces.

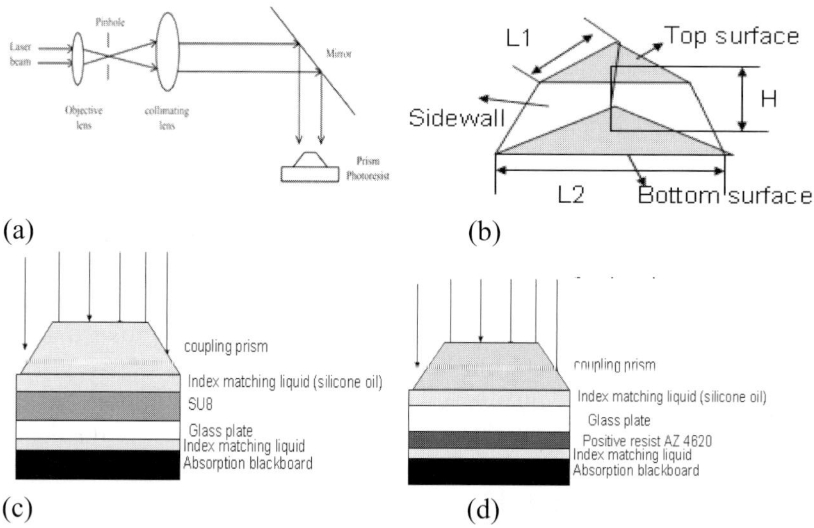

Figure 13.1 (a) A schematic of the holographic lithography setup. The collimated laser beam is reflected by a mirror towards the prism and the glass substrate. (b) The coupling top-cut prism has a symmetric structure. The top and bottom surfaces are equilateral triangles with edges of L1 and L2 respectively. The cutting angle determines the final refracted beam position and can be calculated by $\tan(\phi) = H \times \sqrt{3}/[2(L_2 - L_1)]$. (c) A schematic of the positions of the prism, glass substrate and the absorption plate for the use of negative photoresist. (d) A schematic of the positions of the prism, glass substrate and the absorption plate for the use of positive photoresist.

To model the fabrication of the 3D holographic lithography process more accurately we need to consider the absorption of the laser beam inside the photoresist. With reduced size, the reliability of the actual

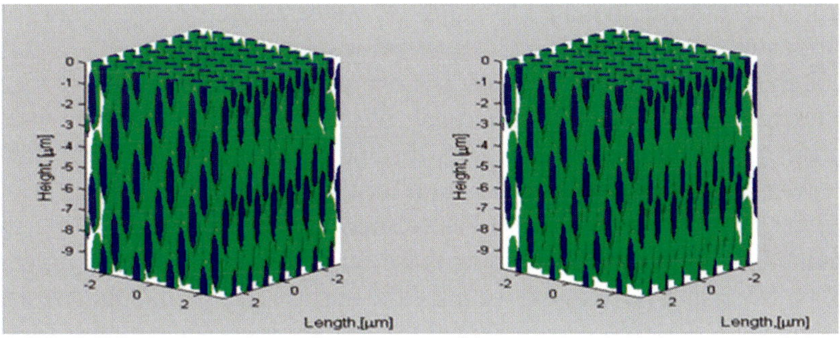

(a) (b)

Figure 13.2 (a) Simulation of intensity distribution in the photoresist without considering absorption. (b) Simulation of intensity distribution in the photoresist considering radiation absorption.

fabricated structures would be more influenced by such absorption. Figure 13.2 shows the MATLAB simulation results of FCC type photonic crystal obtained from four-beam interference with and without absorption using the K vectors obtained before and an SU8 absorption of 0.1/µm at 330 nm. The polarization and the intensity of the beams were calculated using refraction theory at each interface.

The blue volume enclosed by the green surface has an exposure dosage above the polymerization threshold and it would be insoluble during development for a negative resist. It is not trivial for a negative resist 3D photonic crystal with a small feature size structures to survive the development owing to the potentially violent capillary force on a microscopic scale. It is critical to ensure good substrate adhesion and to avoid the structural damage originating from the capillary drainage of the rinsing liquid during the drying process. From the simulation results accounting for the absorption, it is obvious that the bottom part of the resist receives less effective dosage. Therefore, less SU8 resist would remain at the bottom part after development and it is highly susceptible to damage during the drying process. The standard SU-8 rinsing liquid is isopropanol, and it has a surface tension γ of 23 mN/m and a contact angle of $\theta = 20°$. On the contrary, when considering the wetting characteristic, although DI Water has a higher surface tension of 73 mN/m, it has a contact angle of 81° due to the hydrophobic nature of the SU8 material[7]. As the capillary force is proportional to $\gamma\cos(\theta)$, the capillary force exerted by water is about two times smaller than that by isopropanol and we switched to using DI water as the rinsing liquid for all fabricated samples and then no peel-off was observed.

In the case of a positive resist, the exposed blue volume in Fig. 13.2 will be removed during development. With the absorption there is a decrease of dissolution volume with the increased depth of photoresist. Moreover, the actual development time for the upper resist is longer than the lower part as the developer has to etch the upper layers at first to reach the bottom layers. Combining these two effects in the development process, we would have overdeveloped the surface part of the photoresist and damaged the top structures before the desired 3D structure was formed at the lower part of the photoresist. One way to avoid such a problem is to flip the recording glass plate and let the multiple laser beams incident on the glass substrate at first then go into the material instead of direct incidence on the surface of the photoresist. In this way, we have reversed the intensity distribution in the photoresist and during the development the developer would proceed relatively slowly into the lower part of the material then the lower dosage of the surface part would compensate for the longer etch time and maintain the surface structures. The position of the prim and the glass substrate is illustrated in Figure 13.1(d).

In the experiments we used photoresist coated on a glass substrate for the recording of the 3D structures. The negative resist we used was a commercially available SU-8 2007 from Microchem Inc. After the spin coating process, the substrate was baked at 95°C to evaporate the solvent and we obtained a resist thickness of about 10 μm. The 325 nm line from the He-Cd CW laser was used to expose the sample and the light intensity was about 5 mW/cm^2. The index matching liquid employed was silicone oil which does not react with SU8. After the exposure, we applied a post-exposure bake at 95°C to accelerate acid diffusion and induce cationic polymerization of the epoxy groups in SU8. The unexposed film was removed by PGMEA and rinsed in DI water. For the positive photoresist, we used AZ 4620 from Clariant International Ltd. The spin coated film was about 8 μm thick. After the 100°C bake the 442 nm line from the He-Cd laser was used to expose the sample in a flipped fashion mentioned earlier with a light intensity of 10 mW/cm^2. We used the same index matching liquid as in the negative photoresist case. No post-exposure bake was necessary for this AZ 4620 photoresist.

13.3 Characterization of Fabricated Devices

Figure 13.3 shows the SEM images of the fabricated 13.3D polymer photonic crystal structures. The total area is decided by the size of the prism and in our case it is an equilateral triangle with an edge of 1cm. Figure 13.3(a) exhibits the whole size and a fracture of the 3D FCC-type structure using SU8. The detailed fine structures in the (111) plane is presented in Figure 13.3(b) with good uniformity. The enlarged side view of the cleaved SU8 sample with 6 layers is shown in Figure 13.3(c).

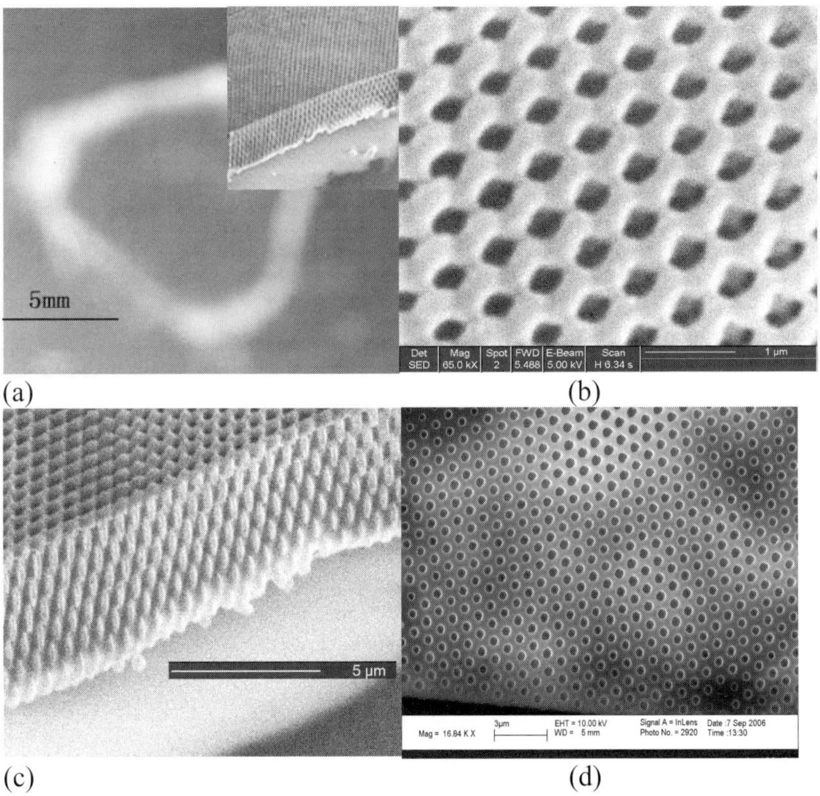

Figure 13.3 (a) The final size of a 3D photonic crystal on SU8 with a 1.1 s exposure. The inset shows a fracture of the cleaved photonic crystal. (b) SEM image of the (111) plane of a 3D photonic crystal on SU8 with 1.1 s exposure. (c) SEM image of a side view of a 3D photonic crystal on SU8 with 1.1 s exposure. (d) SEM image of the (111) plane structure of AZ 4620 with 10 s exposure.

Figure 13.3(d) shows the structure on (111) plane of the AZ 4620 positive resist. For the structures fabricated in the SU8 resist, the (111) in-plane and perpendicular lattice spacings were measured to be 0.61 and 2.02 μm, respectively. The simulation results showed (111) in-plane and perpendicular lattice spacing to be 0.63 and 2.10 μm which are slightly larger than the fabrication results. Such differences can be attributed to the size error of the prism and the shrinkage of the photoresist after the bake process. As for the AZ4620 based structures, the lattice spacing in the (111) plane is about 0.82 μm which is just 1.37 times the one we get in the SU8 case, and such a ratio is consistent with the ratio of the laser wavelengths we used to expose the two different materials. From these observations we have shown that with careful modification to the conventional holographic fabrication process we can obtain a generic approach for fabricating submicron 3D photonic crystals using positive and negative photoresists. In addition, we used the 325 nm laser line normally incident on the (111) plane of 3D photonic crystals in both materials to observe the diffraction patterns which is equivalent to x-ray diffraction normal to the (111) plane of an FCC single crystal[8]. The diffractions from photonic crystals in both materials exhibit similar patterns as shown in Figure 13.4 except for different diffraction angles of 35° and 23° due to the different periodicities. The (111) plane periods calculated from the measured diffraction angles are in good agreement with the SEM measurements and such diffraction patterns further assure us of the formation of good 3D periodic microstructures.

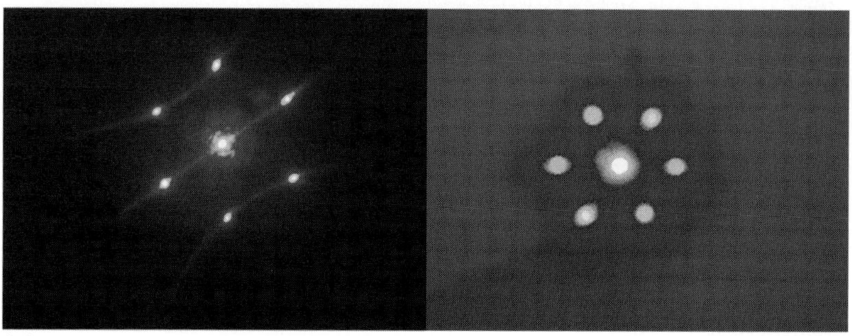

(a) (b)

Figure 13.4 (a) The 1st order diffraction pattern of 325 nm laser line directly incident on the (111) plane of the 3D FCC-type SU8 structure. (b) The 1st order diffraction pattern of 325 nm laser line directly incident on the (111) plane of the 3D FCC-type AZ4620 structure.

We have characterized the optical transmission and reflection properties of the SU8 photonic crystals with an estimated filling factor of 45% using a Fourier transform infrared spectrometer (Thermo Mattson Infinity Gold FTIR). The surface of the sample, which was the (111) plane, was aligned perpendicular to the optical axis in the setup. As shown in Figure 13.5, the dip in the transmission and the peak in the reflection spectrum both occur around 1.5 μm and the band covers 1460-1560 nm. In addition, the bulk SU8 transmission is almost 100% in the whole spectrum range and it is an indication that the fine structures in the spectra are not related to the absorption of the SU8 material. Compared with the band structure in the [111] propagation direction, calculated by plane wave expansion, the position of the dip in the transmission is in good agreement with the simulation with the structural data extracted from the SEM observations. Due to the roughness of the fabricated structures on a 10-nm scale, we have also observed a decrease of the transmission towards shorter wavelengths.

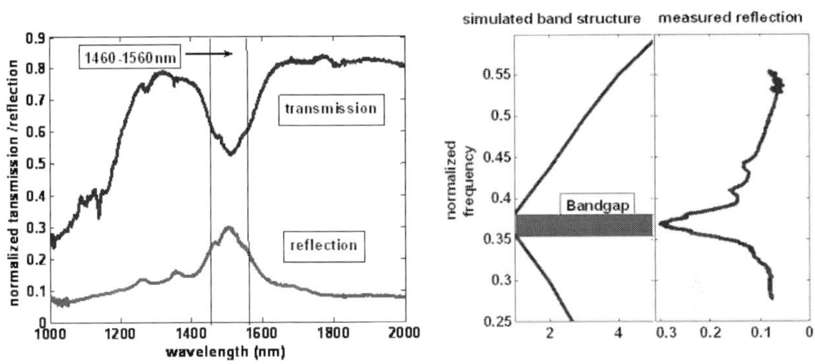

(a) (b)

Figure 13.5 (a) FTIR transmission and refection spectra in the [111] direction. (b) Comparison of the calculated band and the measured spectrum showed good agreement for S-band (1460-1530nm) and C-band (1530-1565nm).

13.4 Superprism Effects Simulation

Superprism refers to the anomalous refraction of light at an interface between a photonic crystal and a homogeneous medium. The refraction angle is found to be very sensitive to the change of incident angle and wavelength under proper conditions. Such an effect arises from the anisotropy of the bands in the photonic crystal and such dispersion effects could be hundreds of times stronger than the conventional prism. And that is where it gains the name "superprism". A number of groups have previously designed and fabricated superprism devices since its introduction by Kosaka and co-workers[9].

Superprism effects in two-dimensional periodic systems were investigated by Baba et al.[10] and then by Chung et al.[11] The first experimental demonstration of 2D superprism effect was reported by Wu et al.[12] who employed an asymmetric GaAs-AlGaAs heterostructure to provide light confinement in the third dimension via total internal reflection. On the other hand, by working with 3D photonic crystals one can get a better control of light in all directions and can achieve a more versatile design. Also, 3D photonic crystals can offer a real 3D superprism effect instead of the in-plane superprism effect.

With the structure data extracted from the previous simulation and experiments, we calculated the (111) in-plane superprism effects. The 3 unit vectors in the real space we used in the simulation were $a_1 = 2.6850\vec{z}$, $a_2 = 0.3084\vec{x}\text{-}0.1780\vec{y} + 0.8950\vec{z}$, $a_3 = 0.3561\vec{y} + 0.8950\vec{z}$ where all the units were in μm. And the corresponding unit vectors for the reciprocal lattice were $b_1 = -10.1879\vec{x}\text{-}5.8820\vec{y} + 2.3401\vec{z}$, $b_2 = -20.3759\vec{x}$, $b_3 = 10.1879\vec{x} + 147.6460\vec{y}$ where all the units were in μm^{-1}.The first Brillouin zone in the x-y plane which is the (111) plane was determined from the in-plane components of the three unit vectors of the reciprocal lattice. It just resembled a 2D triangular lattice in real space with effective lattice spacing of 356.1 nm.

The dielectric constant of the polymer background was taken as 2.56, which was roughly the value for the SU8 resist at the wavelength of interest. The filling factor of the porous 3D structure used in the simulation was 0.45, as determined before. The dispersion curves of the first band in the (111) plane are shown in Fig. 13.6 (a) and (b).

From the above figures, we have observed that around 1550 nm wavelength the dispersion curves exhibit strong anisotropic properties and in that operation region would give us the desired superprism effects. Furthermore, as the off-plane component is small so we can focus on the in-plane propagation in the photonic crystal. In our design we cleaved the sample along the y-axis, and then due to the conservation of $K_\perp$ we can draw a constant K_y line to find the intersections on the dispersion curves and determine the group velocity direction of the transmitted wave. From Fig. 13.6(a) for 1550 nm wavelength there are two equivalent frequency circles having intersections with the corresponding constant K_y line. Thus we needed to decide which intersection point dominates the propagation properties. We investigated the transmission efficiency of each branch which was just how much of the incident light can be coupled to the propagation light determined from each intersection. It turned out that

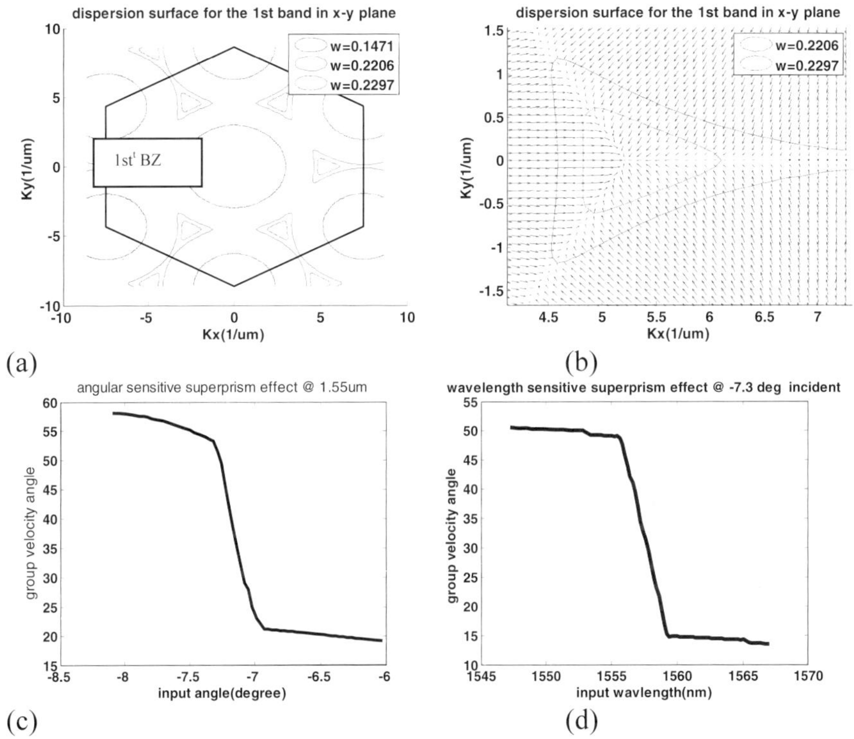

Figure 13.6 (a) The (111) plane dispersion curves for the 1st band of the fabricated 3D polymer photonic crystal. The data curves correspond to wavelengths of 2421 nm, 1614 nm and 1550 nm respectively. The dark color hexagon is the 1st Brillouin zone in the xy-plane. (b) A close look at the "high" transmission part of the dispersion surface when the incident interface is along the y-axis. The arrows indicate the group velocity direction inside the photonic crystal. (c) Angle-sensitive superprism effect at 1550 nm. The negative angle means the incident wave is in the opposite direction with respect to the y-axis and has a negative K_y value. The angle is defined as $\tan^{-1}(-K_x/K_y)$ (d) Wavelength- sensitive superprism effect around 1550 nm.

the one with positive K_x had a transmission larger than 50% while the transmission efficiency for the one with negative K_x was always lower than 10%. From these observations the curves on the positive K_x part were what we were interested in. The angle-sensitive superprism effect at 1550 nm wavelength and the wavelength-sensitive superprism effects at -7.3° incident angle were calculated from the dispersion curves and are shown in Fig. 13.6 (c) and (d).

When the incident angle changed from -7.5° to -7°, the propagation angle inside the photonic crystal changed from 56° to 21° . And when the wavelength changed from 1555nm to 1559nm, there was a dramatic change of the propagation angle from 49° to 15°. Therefore we have achieved superprism effect in 3D polymer photonic crystal structures around the wavelength of 1550 nm, which is suitable for telecommunication use.

References

1. J. D. Joannopoulos, R.D. Meade, and J.N. Winn, Photonic Crystals (Princeton, New York, 1995).

2. W. Jiang, R. T. Chen, and X. Lu, Theory of light refraction at the surface of a photonic crystal, Phys. Rev. B. 71, 245115 (2005).

3. J. Chen, W. Jiang, X. Chen, L. Wang, S. Zhang, and R.T. Chen, Holographic three-dimensional polymeric photonic crystals operating in the 1550 nm window, Appl. Phys. Lett. 90, 093102 (2007).

4. M. Campbell, D. N. Sharp, M. T. Harrison, R. G. Denning and A. J. Turberfield, Fabrication of photonic crystals for the visible spectrum by holographic lithography, Nature (London) 404, 53 (2000).

5. I. Divliansky, T. S. Mayer, K. S. Holliday and V. H. Crespi, Fabrication of three-dimensional polymer photonic crystal structures using single diffraction element interference lithography, Appl. Phys. Lett. 82, 1667 (2003).

6. L. Wu, Y. Zhong, C. T. Chan, K. S. Wong and G. P. Wang, Fabrication of large area two- and three-dimensional polymer photonic crystals using single refracting prism holographic lithography, Appl. Phys. Lett. 86, 241102 (2005).

7. T. Kondo, S. Juodkazis, and H. Misawa, Reduction of capillary force for high-aspect ratio nanofabrication, Appl. Phys. A. 81, 1583–1586 (2005).
8. N. W. Aschcroft and N. D. Mermin, Solid State Physics (Philadelphia : Saunders College,1976).
9. H. Kosaka, T. Kawashima, A. Tomita, M. Notomi and T. Tamamura and T. Sato and S. Kawakami, Superprism phenomena in photonic crystals, Phys. Rev. B. 58, 10096 (1998).
10. T. Baba and M. Nakamura, Photonic crystal light deflection devices using the superprismeffect, J. Quantum Electron, 38, 909 (2002).
11. K. B. Chung and S. W. Hong, Wavelength demultiplexers based on the superprism phenomena in photonic crystals, Appl. Phys. Lett. 81, 1549 (2002).
12. L. Wu, M. Mazilu, T.F. Krauss, Beam steering in planar-photonic crystals: from superprism to supercollimator, J. Lightwave Technology. 21,561, 2003.

14

CONTINUOUS ROLL NANO-IMPRINTING TECHNOLOGY FOR LARGE-SCALE NANO- AND MICROSTRUCTURES

Shinill Kang, Suho Ann, Seok-min Kim

A continuous roll nano-imprinting process is regarded as one of the most promising technologies for realizing large-scale nano- and microstructures at low cost. This technique is applicable to mass production of devices in fields such as digital displays, nano energy, nano data storage etc. This chapter explains some fabrication methods for making roll stamps and key issues in thermal and UV roll nano-imprinting processes.

14.1 Introduction

The increasing demand for nano- and microstructures on large substrates makes the establishment of large-scale fabrication technology for polymeric nano- and microstructures a priority in the fields of digital display, nano energy, nano data storage etc. as exemplified in Figure 14.1. Several technologies are considered promising for low-cost mass production of nano- and microstructures, including injection molding, hot embossing, and UV imprinting[1-6]. Injection molding has been regarded as one of the best methods, with this automated process used to

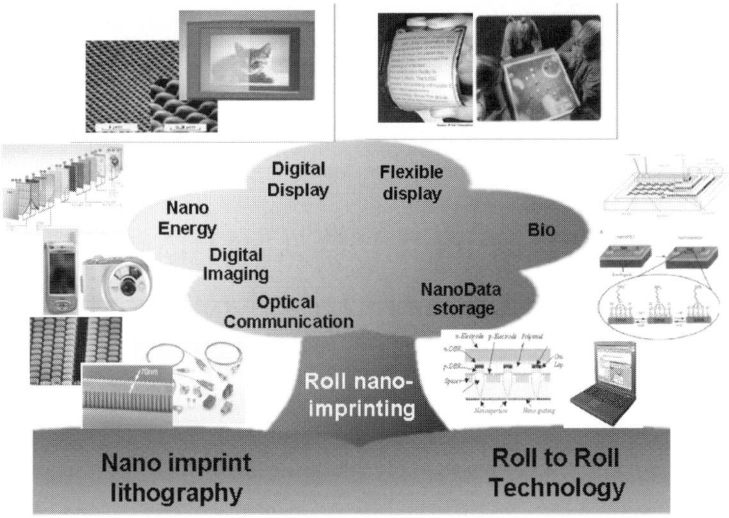

Fig. 14.1 Schematics of application fields of roll nano-imprinting process.

replicate optical data storage media (CD, DVD, and Blu-ray substrates) and to produce components with nano- and microstructures (the light-guide plate of liquid-crystal displays(LCD))[1,2]. Thermal imprinting involves shaping the thermoplastic substrate after it has been softened above its glass transition temperature, and is another mass-production method for fabricating polymeric nano- and microstructures in applications such as photonic crystals and optical wave guides[3,4]. UV imprinting involves applying a stamp to a liquid resin and then exposing it to UV light for curing and is regarded as the most suitable process for fabricating or integrating nano- and microstructures, including sub-wavelength optical elements, diffractive optical elements, microlens arrays on vertical cavity surface emitting lasers (VCSELs), and microlens-integrated image sensor[5,6], because the process does not require high pressures since the resin is basically a viscous liquid that is soft enough to be easily deformed.

Fabricating nano- and microstructures for commercial devices requires the development of large-scale rapid fabrication technologies[7]. The size of devices is especially important in the display market, which is one of the largest markets for electronic and photonic devices (including functional optical films). Furthermore, the ability to fabricate large-scale nano- and microstructures would offer economies of scale since it would allow manufacturers to produce a greater number of devices from a single fabricated structure. Conventional replication processes usually employ a flat stamp, and hence developing a large-scale flat stamp is a natural starting point in efforts to develop a large-scale replication process. However, machining large-scale flat stamps is very difficult, since the machining costs increase exponentially with the stamp size[8]. Furthermore, previous efforts to use a replication process to fabricate large-scale nano- and microstructures have encountered uniformity and release problems.

Roll-printing technologies such as gravure offset printing and flexography printing offer alternative approaches to large-scale structure fabrication[9-10]. Although roll-printing technology has the advantages of being cost effective and providing high-throughput processing, the pattern widths of roll-printing processes have generally been limited to tens of micrometers.

To produce large-scale three-dimensional nano- and microstructures with high pattern fidelity and throughput, a roll nano-imprinting process is proposed. In the roll nano-imprinting process, the cavities on the roll stamp was transferred onto the imprinting material, which provides several advantages such as good release, easier control of defects, and high uniformity[11].

14.2 Fabrication of Roll Stamp

The roll nano-imprinting process is crucially dependent on the fabrication of the roll stamp that contains nano- and microstructure cavities because the dimensional accuracy and uniformity of the roll stamp and cavities determine the properties of the roll nano-imprinted parts. Various methods can be used to fabricate the roll stamp for the nano- and microstructures, including direct mechanical machining of the roll base and wrapping a thin metal or polymer stamp around the roll base.

14.2.1 Direct mechanical machining

If the required cavity is larger than about 50 μm and the cavity is one-dimensional, such as in a lenticular lens array for a flat panel display with a wide viewing angle or a prism sheet for enhancing the brightness of an LCD backlight unit then a conventional computer numerically controlled machine using a roll base as a work piece can be used to fabricate the roll stamp. Various stamp materials can be used in this process, including metal, silicon, glass, polymer, and tungsten carbide, although only materials with good machinability (e.g., nickel, aluminum, and copper) are suitable if the required minimum features size is less than about 10 μm[12]. A shaped single diamond tool is normally used to fabricate roll stamps that have microcavities with sharp corner or small angles. Direct mechanical machining of the roll base is a simple process that can produce highly accurate microcavities without any joining line on the roll stamp. Figure 14.2(a) and (b) are images of the tooling process and a machined aluminum roll stamp with lenticular lens cavities, respectively[13]. In the direct machining of the roll stamp, the work piece was rotated along the longitudinal axis of the roll base and the cavities were machined by a shaped single diamond tip. However, direct machining has some limitations when microstructures are required along the axial direction of the roll base or where very small cavities are required.

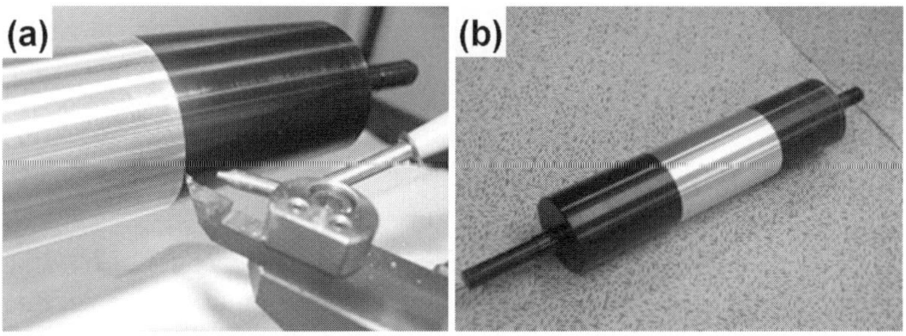

Fig. 14.2 Images of (a) the tooling process and (b) a machined aluminum roll stamp with a length of 210 mm and a diameter of 50 mm[13].

14.2.2 Wrapping of a thin metal or polymer stamp onto a roll base

If one can wrap a thin metal stamp or a polymer stamp around the roll base, it can be used to produce a roll stamp with cavities of various shapes, including for microlens arrays, pyramidal pits, and nanoscale structures. Various nano- and micropatterning processes can be used to fabricate a thin metal or polymer stamp. These include, direct machining with nanometer accuracy including diamond-tool shaping, laser ablation, focused-ion-beam machining, and electroforming or polymer replication of a master with nano- and microstructures. The electroforming or polymer replication of a master structure is the most effective technique for a thin metal or polymer stamp which contains arbitrary continuous nano- and microscale surface-relief profiles. The master structure can be fabricated using methods such as electron-beam lithography, laser interference lithography, photolithography, reflow, etching etc.[12]

Various methods can be used to wrap a thin stamp with preformed nano- and microstructure cavities around the roll base. One simple method is to adhere the stamp uniformly to the peripheral face of a mirror roller (a roller with mirror-finished surface) using an adhesive, as depicted in Figure 14.3(a)[13]. For this method, adhesive durability is a very important consideration since the heat and pressure of the roll nano-imprinting process make it liable to peel off from the mirror roller. Bubbles trapped between the mirror roller and the thin stamp must also be avoided since they would interfere with the precise transfer of the structures, resulting in defects. Finally, the adhesive must have a uniform thickness.

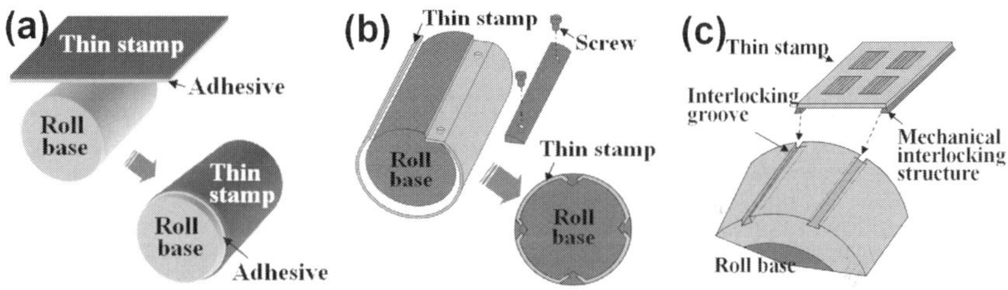

Fig. 14.3 Schematic diagrams of the processes of wrapping a thin stamp to the mirror roller using (a) adhesive[13], (b) screws[14], and (c) mechanical interlocking structures[15].

Another method involves mechanical fastening of the thin metal or polymer stamp to the mirror roller[14,15]. Figures 14.3(b) and (c) are schematic diagrams of examples to demonstrate mechanical fastening of a thin stamp to the roll base using screws and a mechanical interlocking structure, respectively. When using screws, the jig shape and the screw positions are carefully designed to avoid unevenness or distortion at the fixation points, since these may result in a defective pattern transfer or exert an uneven pressure on the imprinted sheet, resulting in, for example, birefringence of the roll nano-imprinted devices. When using a mechanical interlocking structure, the structure has a trapezoid section that is bonded to the back of the stamp along each edge parallel to the roller axis. The mirror roller contains interlocking grooves at its peripheral face with an identical cross-section for receiving the interlocking structure. This latter method is dependent on the development of a method for machining or bonding the mechanical interlocking structure and the interlocking grooves.

14.3 Continuous Thermal Roll Imprinting

Continuous roll nano-imprinting process is highly suitable for mass fabrication of nano- and microstructures. Some commercial products have been produced using a continuous thermal roll nano-imprinting process, including security features with a holographic nanostructure[4].

In the conventional flat-type thermal imprinting process, the cavity structure is directly transferred onto a thermoplastic substrate that is heated to above its glass transition temperature, at which point the substrate becomes viscous and conforms exactly to the stamp by filling the cavities of the surface relief. After it has cooled down, the replica is released from the stamp. The long cycle time associated with the heating and cooling processes and its batch-wise nature make conventional thermal imprinting a less efficient method for mass production. In a continuous thermal roll nano-imprinting process, a structured roll stamp is pressed into a thermoplastic polymer layer. Although a metal roll stamp is normally used due to its good durability in a high-temperature environment, a polymer stamp can also be used. As the cylinder rotates, the cavities of the roll stamp are continuously pressed into the thermoplastic sheet that is fed into the gap between two cylinders. The sheet and stamp are not in thermal equilibrium throughout the molding sequence; instead, the temperature of the polymer rises locally above the glass transition temperature only when its surface is in contact with the hot solid roll stamp. This eliminates the time-consuming heating and cooling cycle in conventional flat-type thermal imprinting, thereby increasing productivity. The thermal roll nano-imprinting process has other advantages in eliminating the defects that are usually observed in thermal imprinting processes using a flat stamp. In the thermal roll imprinting process, microscopic air bubbles can easily be eliminated without evacuation process because the roller pushes the air out. During flat thermal imprinting, the entire area of the sample is imprinted at the same time, and hence unevenness in stamp thickness or the presence of even a single small dust particle can affect the uniformity over a large surrounding area. In contrast, in thermal nanoscale roll imprinting the contact occurs only in one dimension at a given time, significantly reducing the effects of thickness variations and dust in surrounding areas. Makela et al.[16,17] developed a custom-made continuous roll nano-imprinting system in which the roll stamp is heated above the glass transition temperature of the polymer using heating elements inside the roll. The system operates at a maximum of 200°C, 20 m/min, and 1250 N/cm. Figure 14.4(a) shows a photograph of the system, in which thin nickel stamps were attached to the printing cylinder mechanically or using adhesive tape. Various nano- and microstructures have been produced using this system. Chou et al.[18] suggested two types of thermal roll nano-imprinting process: (1) using a roll stamp fabricated by wrapping a thin metal stamp around a smooth roller, and (2) using a flat mold placed directly on a substrate. In the second process a smooth roller rotates over the mold and the slight deformation of the flat mold under the pressure of the roller imprints structures in the substrate. A smooth roller with a lamp-based heater used in two thermal roll nano-imprinting processes can produce a temperature of up to 200°C. The feed speed is 0.5–1.5 cm/min and the pressure is 300–4800 psi. Determining the optimum temperature, pressure, and speed for various combinations of mold material, molding material, and cavity size revealed that the imprint quality was poor when the temperature of the roll base or the feed speed was too high, or when the pressure was too low. They successfully fabricated a one-dimensional grating with a pitch of 190 nm and a depth of 40 nm, as depicted in Figure 14.5(a).

In the thermal roll nano-imprinting process, the heated substrate should be cooled to below its grass transition temperature before being released from the roll stamp. However, the short contact time usually means that the released structures on the substrate is still viscous, resulting in a poor imprint quality. Seo et al.[19] developed a roll nano-imprinting system in which a rigiflex stamp is in contact with the substrate after the roll pressing process, with it being released from the substrate only once it has sufficiently cooled. Figure 14.4(b) shows the schematics of a thermal roll nano-imprinting system using a poly(urethaneacrylate) material (PUA) as a rigiflex stamp, which is sufficiently rigid for sub-100-nm structures while also being sufficiently flexible. The system consists of two rollers for heating and two rollers for releasing. A heating wire is used to externally heat the substrate as it comes into contact with the bottom roller. The upper rollers are coated with poly(dimethylsiloxane) (PDMS), which provides a consistent contact with the polymer sheet stamp and a uniform pressure distribution, with the high friction provided by the PDMS also preventing the polymer stamp from slipping from the roller. Figure 14.5(b) shows an SEM image of an imprinted polystyrene structure with a line width of 500 nm and a depth of 500 nm obtained using a continuous rigiflex imprinting process.

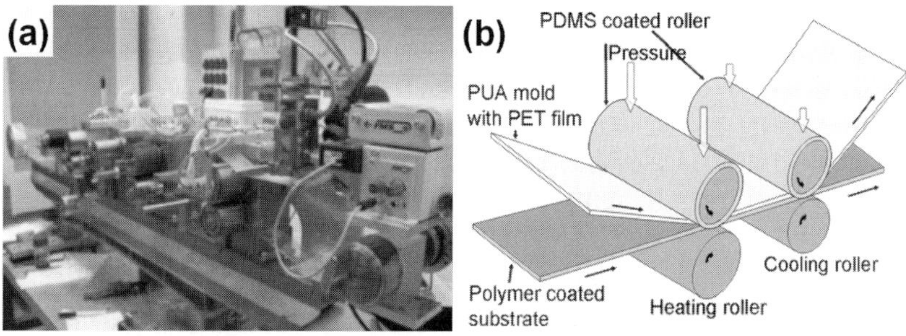

Fig. 14.4 (a) Photograph of a thermal roll nano-imprinting system using a roll stamp[17], and (b) schematic of a continuous rigiflex imprinting process[19].

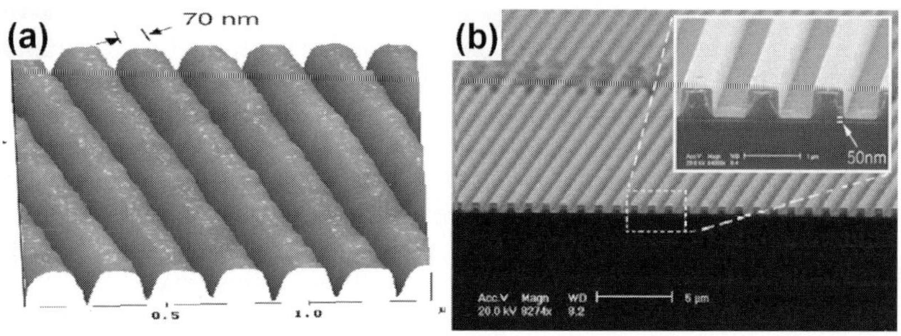

Fig. 14.5 (a) AFM measurement results of a thermal roll nano-imprinted one-dimensional grating with a pitch of 190 nm and a depth of 40 nm[18]. (b) SEM image of an imprinted polystyrene structure with a line width of 500 nm and a depth of 500 nm[19].

14.4 Continuous UV Roll Imprinting

A low-pressure and low-temperature UV imprinting process has been developed to significantly reduce the heating and cooling times of the thermal imprinting process. The process employs UV light to cure a UV-curable photopolymer. To increase the productivity and scale up to large surface areas, a continuous UV roll imprinting process was suggested[13,20-22]. Figure 14.6 shows schematic diagrams of the continuous UV roll nano-imprinting system for rigid and flexible substrates, which has the following components: the material coating unit, roll stamp, contact roller, releasing roller, and exposure unit[20]. The substrate, which has been coated with a UV-curable photopolymer, is passed between the roll stamp and the contact roller. In the contact region between the roll stamp and the substrate, the photopolymer fills the cavities in the roll stamp, and the resin is simultaneously cured by UV irradiation from the UV-exposure system.

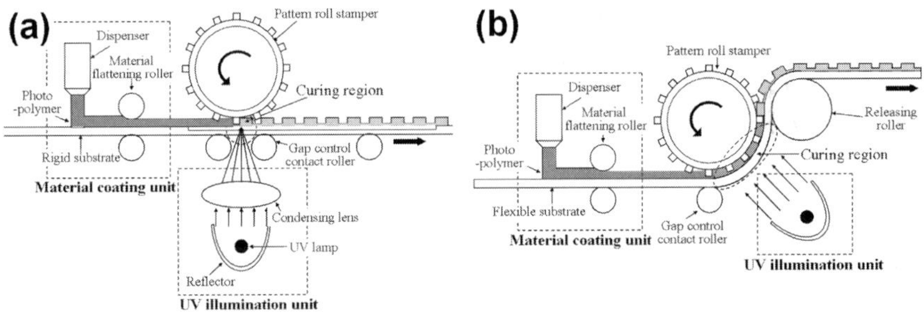

Fig. 14.6 Schematic diagrams of a UV roll nano-imprinting system for a rigid substrate (a) and a flexible substrate (b)[20].

The main issues to be addressed in the continuous UV roll nano-imprinting process is the attainment of high uniformity of material coating onto the substrate, controlling the irradiated energy to the photopolymer, obtaining a residual layer with a uniform thickness and good fidelity of imprinted structures. Since at room temperature the UV photopolymer is a viscous material just like adhesive and ink, standard methods used in the roll-printing and adhesive-coating industries can be used to coat the UV photopolymer onto the substrate, such as direct or reverse gravure, reverse roll, die coaters, wire bar, knife or blade coaters etc. A microgravure coating method for precise control of the amount of coating material has also been developed in which the gravure roll surface is engraved with cells that provide a specific coating volume[23]. The microgravure roll picks the material on the roll surface including the cells and then the coated material (except that in the cells) is removed by a flexible steel blade, leaving the exact amount of material in the cells to be coated on the substrate.

Curing the UV photopolymer requires sufficient exposure dose of UV light which can be controlled by the intensity of the irradiated UV radiation, the length of the exposure region, and the roll nano-imprinting speed. Optimizing the roll nano-imprinting speed requires the effect of the exposure dose on the curing state of the photopolymer to be analyzed. Among the various methods used to measure the degree of photopolymerization, Fourier-transform infrared (FTIR) spectroscopy has often been used to determine the chemical structure through the detection of the absorption or scattering of infrared light. During the polymerization of the photopolymer, the carbon double bond in the monomer state is converted into a carbon single bond. In the monomer state, a photopolymer (urethane acrylate) has absorption peaks in its FTIR spectrum at 3037 and 3100 cm^{-1} due to the carbon double bond, as shown in Figure 14.7[20].

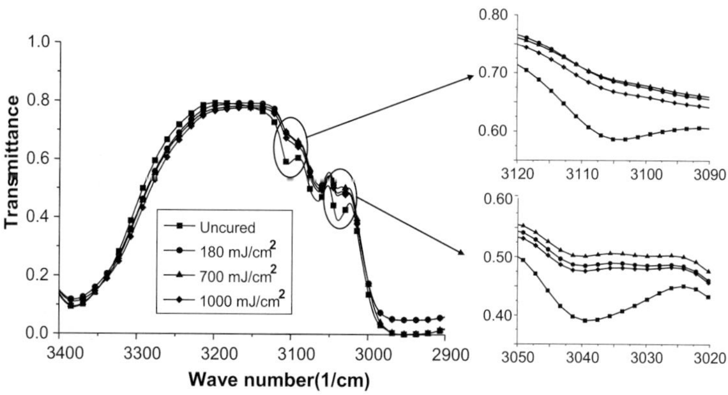

Fig. 14.7 Data from Fourier-transform infrared (FTIR) spectroscopy of urethane acrylate resin. In the monomer state, urethane acrylate has absorption peaks in its FTIR spectrum at 3037 and 3100 cm^{-1} due to its carbon double bonds[20].

Obtaining acceptable thickness uniformity and replication quality requires a near-zero residual layer (since variation in the residual layer thickness may result in poor thickness uniformity), and this requires a very small gap between the roll stamp and the substrate. A pre curing process was considered for ensuring thickness uniformity[24], but controlling the pre-exposure UV light dose is difficult, and undesirable degradation of properties (e.g., birefringence and low fidelity) may occur in the roll nano-imprinted structures. A passive gap-control system was suggested for controlling the thickness uniformity of the residual layer and increasing the fidelity of imprinted parts[20]. Ball-spring plungers that press the contact roller against the roll stamp can be used as a passive gap controller.

Various nano- and microstructures were replicated using the UV roll nano-imprinting system. Figure 14.8 shows SEM images of UV roll nano-imprinted nano- and microstructures, demonstrating their successful fabrication with high fidelity, including high-aspect-ratio nanostructures and microstructures with sharp edges. The roll stamps for one-dimensional grating, nanopillar array, and pyramidal pits were fabricated by wrapping thin electroformed nickel stamps onto the roll base, with the masters fabricated by laser interference lithography, electron-beam lithography, and mechanical machining, respectively.

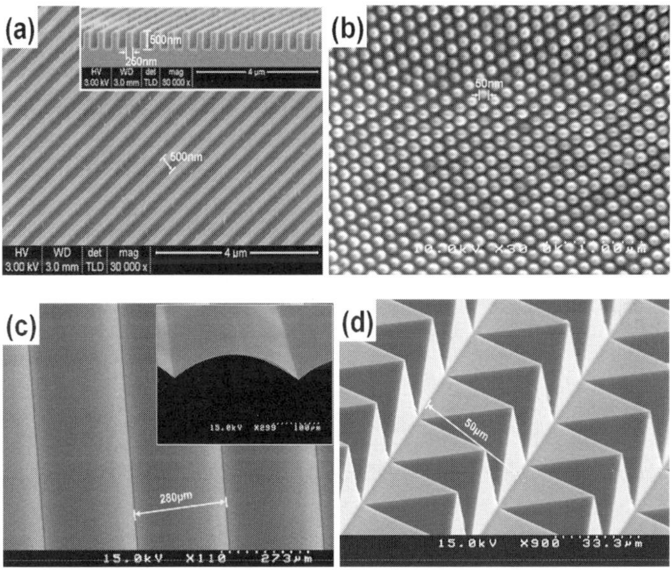

Fig. 14.8 SEM images of UV roll nano-imprinted nano- and microstructures: (a) one-dimensional grating with a pitch of 500 nm, a height of 500 nm, and an aspect ratio of 2, (b) nanopillar array with a diameter of 50 nm, a pitch of 150 nm, and a height of 35 nm[20], (c) lenticular lens array with a pitch of 280 μm and a sag height of 47 μm[20], and (d) pyramidal pits with a pitch of 50 μm and a height of 25 μm[13].

14.5 Conclusion

The roll nano-imprinting process shows considerable promise for the mass production of nano- and microstructures on large substrates. It avoids many of the problems that are characteristic of conventional flat nanoscale imprinting processes, including microscopic air bubbles, unevenness of the residual layer, and defect formation during the releasing process. The roll stamp can be fabricated by a direct machining process for simple microscale structures and by wrapping a thin stamp for complex micro- and nanoscale structures. In the thermal roll nano-imprinting process, the processing temperature, rolling speed, and contact force are the dominant factors affecting the properties of the imprinted part. In the UV roll nano-imprinting process, the amount and uniformity of the initially coated material, UV irradiation energy, and contact force are the dominant factors. Some nano- and microstructures have recently been fabricated by various researchers. These include, security holograms, functional optical films for displays etc. However, some issues in the roll imprinting process for nano- and micropatterns, such as the joining line of thin stamp wrapped roll stamp, slip between roll stamp and the substrate and alignment between the stamp and the substrate need to be solved before the process can be used for real industrial applications that require large-area nano- and microstructures at low cost.

References

1. J. Giboz, T. Copponnex and P. Mele, *Microinjection molding of thermoplastic polymers: a review*, J. Micromech. Microeng. 17, R96 (2007)

2. Y. Kim, J. Bae, H. Kim and S. Kang, *Modelling of passive heating for replication of sub-micron patterns in optical disk substrates*, J. Phys. D: Appl. Phys. 37, 1319 (2004)

3. C. Kee, S. Han, K. B. Yoon, C. Choi, H. K. Sung, S. S. Oh and H. Y. Park, S. Park and H. Schift, *Photonic band gaps and defect modes of polymer photonic crystal slabs*, Appl. Phys. Lett. 86, 051101 (2005)

4. M. Minamiti, Y. Utsumi, and T. Hattori, *3D microstructure fabrication for a high luminosity lighting-panel for LCD using synchrotron radiation*, Microsyst. Technol. 11, 230 (2005)

5. M. T. Gale, C. Gimkiewicz, S. Obi, M. Schnieper, J. Sochtig, H. hiele and S. Westenhofer, *Replication technology for optical microsystems*, Opt. Laser. Eng. 43, 373 (2005)
6. S. Kim, H. Kim and S. Kang, *Development of a UV-imprinting process for integrating a microlens array on an image sensor*, Opt. Lett. 31, 2710 (2006)
7. C. W. Mclaughlin, *Progress in projection and large-area displays*, Proc. IEEE 90, 521 (2002)
8. Y. Mikami, Y. Nagea, Y. Mori, K. Kuwabara, T. Saito, H. Hayama, H. Asada, Y. Akimoto, M. Kobayashi, S. Okazaki, K. Asaka, H. Matsui, K. Nakamura, and E. Kaneko, *A new patterning process concept for large-area transistor circuit fabrication without using an optical mask aligner*, IEEE T. Electron Dev. 41, 306 (1994)
9. M. Pudas, J. Hagberg, and S. Leppavuori, *Printing parameters and ink components affecting ultra-fine-line gravure-offset printing for electronics applications*, J. Eur. Ceram. Soc. 24, 2943 (2004)
10. M. Pudas, J. Hagberg, and S. Leppävuori, *The absorption ink transfer mechanism of gravure offset printing for electronic circuitry*, IEEE T. Compon. Pack. B. 25, 335 (2002)
11. P. Kao, S. Chu, C. Zhan, L. Hsu, W. Liao, *Fabrication of the patterned flexible OLEDs using a combined roller imprinting and photolithography method*, 5th IEEE Conference on Nanotechnology 2, 693 (2005)
12. S. Kang, *Replication technology for micro/nano optical components*, Jpn. J. Appl. Phys. 43, 5706 (2004)
13. H. Bae, M. Choi, S. Ahn, J. Lim, H. Myung, H. Kim, and S. Kang, *Design and fabrication of micro optical film by UV roll imprinting*, Jpn. J. Appl. Phys. 46, (2007)
14. T. Santoh, H. Sugata, M. Yashima, H. Kamitakahara, H. Yoshino, O. Kanome, T. Sato, H. Hayashi, *Process for producing substrate sheet for information recroding mediums*, US Patent 5,516,469 (1996)
15. T. Sato, O. Kanome, H. Yoshino, H. Hayashi, H. Kamitakahara, *Roll-stamper for forming substrate of information-recording medium*, US Patent 5,173,313 (1992)
16. T. Makela, T. Haatainen, P. Majander and J. Ahopelto, *Fabrication of plastic microfluidic channels using roll to roll hot embossing*, 5th International Nanoimprint and Nanoprint Technology Conference, 45 (2006)

17. T. Makela, T. Haatainen, P. Majander and J. Ahopelto, *Roll to roll nanoimprinting on inherently conducting polyaniline*, Proc. Micro- and Nano- Engineering (MNE'06), 6B-5 (2006)
18. H. Tan, A. Gibertson, S. Chou, *Roller nanoimprint lithography*, J. Vac. Sci. Technol. B 16(6), 3926 (1998)
19. S. Seo, T. Kim and H. Lee, *Simple fabrication of nanostructure by continuous rigiflex imprinting*, Microelectron. Eng. 84, 567 (2007)
20. S. Ahn, J. Cha, H. Myung, S. Kim and S. Kang, *Continuous ultra-violet roll nanoimprinting process for replicating large-scale nano- and micro-patterns*, Appl. Phys. Lett. 89, 213101 (2006)
21. T. Makela, T. Haatainen, P. Majander and J. Ahopelto, *Continuous roll to roll nanoimprinting of inherently conducting polyaniline*, Microelectron. Eng. 84, 877 (2007)
22. S. Liu and Y. Chang, *A novel soft-mold roller embossing method for the rapid fabrication of micro-blocks onto glass substrates*, J. Micromech. Microeng. 17, 172 (2007)
23. C. Y. Chang, S. Y. Yang, J. L. Sheh, *A roller embossing process for rapid fabrication of microlens arrays on glass substrates*, Microsyst. Technol. 12, 754 (2006)
24. Y. Kono, A. Sekiguchi, Y. Hirai, S. Arasaki and K. Hattori, *Study on nano imprint lithography by the pre-exposure process (PEP)*, Proc. SPIE 5753, 912 (2005)

15

FABRICATION AND CHARACTERIZATION OF TWO-DIMENSIONAL ZnO PHOTONIC NANOSTRUCTURES

Jingbiao Cui

Periodically ordered ZnO nanopillar arrays were fabricated by a combination of soft templates created by electron beam lithography and an electrochemical process. Growth of ZnO at 90°C in an aqueous solution ensured compatibility with the polymethyl methacrylate used as a template material. Individual ZnO nanopillars with diameters around 100 nm can be precisely placed in desired locations to form two-dimensional periodic structures. Visible light diffraction patterns from the ordered arrays were observed, and the in-plane photonic band-gap was characterized. This approach provides a new method for the design and fabrication of ZnO photonic materials with arbitrary pattern structures and intentional defects. These structures may have potential applications in the manufacture of photonic and optoelectronic devices.

15.1 Introduction

Photonic crystals are artificially constructed structures that constitute highly-ordered, one-, two- or three-dimensional, spatially-aligned objects, in which the dielectric constant is periodically modulated on a length scale comparable to the desired electromagnetic wavelength of operation. These nano/micro-structured materials exhibit a photonic band-gap within which a set of wavelengths are forbidden to propagate in the crystal similar to the electronic band-gap in semiconductors. Photonic crystals have potential applications in photon control and manipulation. They are candidates for inhibiting spontaneous emission, enhancing the performance of semiconductor lasers, providing a new means for guiding light, and playing a fundamental role in the integration and miniaturization of optical components.

Ion-beam-etched structures in silicon have been investigated to prove the principles[1,2] and recently, some progress has been made in the visible region of the spectrum. ZnO is a promising material for such applications due to its wide band-gap and large exciton binding energy. This will be important for many applications such as sensors where higher energy photons are required, for the production of a variety of photonic band-gap devices and for coupled UV laser arrays. Calculations done for ZnO nanorod arrays demonstrated that photonic band-gap structures can be made in these materials in the visible region of the spectrum[3]. However, the fabrication of well-organized two- or three-dimensional vertical ZnO nanorod arrays with controllable structure and period is challenging.

Dense, randomly aligned ZnO nanowire arrays have been obtained using both catalytically activated vapor phase processes at high temperature and hydrothermal solution routes at low temperature[4-9]. Since many applications such as photonic devices require spatial control of individual nanowires, a number of investigations have been devoted to the patterned growth of ZnO nanowire arrays using both vapor phase[10-14] and solution routes[15].

In high temperature vapor phase growth, controlling the size and location of Au catalyst on the substrate surface has been used for the placement of ZnO nanowires. Yang and colleague realized the growth of ZnO nanowires on a photo-lithographically created Au pads on an Al_2O_3 substrate in a vapor phase process[10]. Due to the limitation of photolithography, each patch is of micrometer dimensions and contains tens to hundreds of ZnO nanowires. Both phase shift photolithography and electron beam lithography were used to create ordered arrays of Au pads with size down to 50-200 nm[11,12]. Several nanowires were found to grow on each site without uniformity. Recently, self-assembled monolayers (SAM) of sub-micrometer spheres with a hexagonal close-packed structure were also applied to create Au islands for ordered ZnO nanowire growth[13,14]. However, this technique lacks the capability for designing and creating desired defects in the ordered structures.

The solution routes to ZnO nanowire growth possess numerous advantages including low temperature and little limitation in substrate selection, low cost, and scale-up possibilities. Hsu et al.[15] realized the directed micron-level spatial organization of ZnO nanorods on a silver-coated Si substrate through inhibition of ZnO nucleation using a contact-printed self-assembled monolayer. Most experimental studies demonstrate the deposition of a group of nanowires rather than an individual one on each growth site, although nucleation of individual platelets by the addition of citrate to the growth solution was observed. Recently, Matsuu et al.[16] described the formation of ZnO pillar arrays by filling deep polymer templates and observed photonic band-gap in the visible wavelength. Kim et al.[17] demonstrated localized growth of ZnO columns on ZnO films grown at 95°C. The columns with low aspect ratios were vertically aligned, but for small columns, deviations from the vertical were observed.

The organization of high aspect ratio, vertical nanowires with a controllable pattern structure, intentional defects, and variable periods is a key issue for many applications in photonics such as waveguides. In a recent work, Cui et al.[18] demonstrated the production of two-dimensional ZnO based photonic band-gap crystals as well as line defects within the crystal using a lithographically generated soft template. Measurements of the in-plane reflectance of the nanowire arrays indicate high energy band-gaps in the visible wavelength, which was repeated in the band structure calculations. Importantly, photonic structures with complex patterns were fabricated for the needs of many potential applications. Growth was performed using a reducing potential during a hydrothermal process. Individual single crystal ZnO nanowires were precisely organized using e-beam lithography generated soft template - polymethyl methacrylate (PMMA). This process is compatible with other semiconductor fabrication methods and has the potential for large scale fabrication of two-dimensional photonic structures.

15.2 Experimental Procedures

ZnO photonic structures were fabricated by depositing ZnO nanowires on templated substrates using an electrochemical process, which is a modification of a conventional hydrothermal process by applying a direct-current (DC) electrical potential to the substrate. This new approach showed a number of advantages over hydrothermal growth such as high nucleation density, high growth rate and the possibility of dopant incorporation in ZnO nanowires. The detail of the electrochemical method can be found in literature[8]. Briefly, an aqueous solution (0.006 M) of zinc nitrate and hexamethylenetetramine (also known as hexamine) was heated to 90°C for ZnO nanowire growth. A negative potential was applied to the substrate relative to a gold wire immersed in the solution, to enhance ZnO nanowire growth.

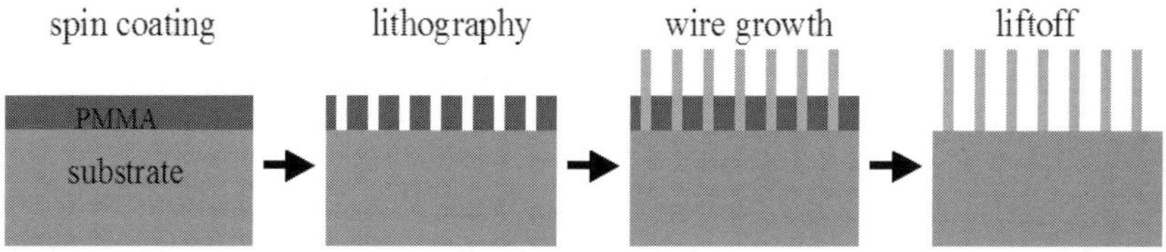

Fig. 15.1 Experimental procedure for ordered ZnO nanopillar growth. The thickness of the PMMA layer is about 200 nm. The diameter of growth windows is between 100 and 200 nm for individual nanowire growth.

PMMA templates were created on Au-coated silicon substrates by e-beam lithography. The procedure for creating the template is shown in Fig. 15.1. A polished silicon (111) substrate was first coated with 2 nm of Cr for adhesion followed by 20 nm of Au and then spin-coated with a PMMA layer. The PMMA, (molecular weight = 495,000; 4% in Anisole) was spun at 4000 rpm to yield a 180 nm thick layer. The resist was exposed using a field emission scanning electron microscope (SEM) and Nabity® software to form a pattern of holes. After a standard development procedure, the substrate with the template holes was ready for use. The templated substrate was then placed in a zinc nitrate solution for ZnO nanowire growth. As shown in Fig. 15.1, the ZnO nanopillars grow on the open areas where the substrate surface is exposed to the growth solution. The PMMA template is unaffected by the aqueous solution at low temperature. Ordered nanopillar arrays were obtained after a liftoff process using remover PG (MicroChem). Large area patterns of nanopillars could be produced by an electrochemically assisted growth. Note that the low growth temperature and mild conditions in the aqueous solution ensure the survival of the PMMA template. Although some ZnO grows on the PMMA as well as in the wells, these structures were removed by a conventional liftoff process after growth. Therefore, only the periodically ordered ZnO nanowire arrays were left on the silicon substrate.

It is noted that the conventional hydrothermal deposition was found to be unsuitable for the growth of ordered nanowire arrays using the PMMA template due to reduced nucleation in the wells. The low nucleation density was found to cause a large number of defects (missing wires) in the smaller windows used for the fabrication of nanopillar arrays. However, the electrochemical potential helped enhance nucleation within the small growth windows and ensured the growth of two-dimensional periodically ordered ZnO nanowire arrays with fewer defects.

The ZnO nanopillar arrays were characterized chemically, structurally and optically. An SEM with energy dispersive x-ray spectroscopy (EDX) was used for morphology and chemical studies. Composition mapping over the sample surface was obtained by scanning the electron beam on the region of interest while a specific detection energy was monitored.

Visible light studies included both diffraction and a measurement of the in-plane reflectivity of the sample. The optical diffraction properties of the ZnO nanowire arrays were investigated with both monochromatic (532 nm) and white light (tungsten lamp). The experimental setup is shown in Fig. 15.2(a). Light from a microscope illuminator was focused on the sample surface by a lens placed in the central hole of a display board. The diffraction pattern was projected onto the display board and recorded by a CCD camera. The separation between the sample and the display board was 15 cm. A digital camera, placed about 50 cm behind the sample, was used to record the diffraction patterns.

Photonic band-gaps can be directly observed in either transmission or reflection. Measurements of the in-plane reflectance were made here using a technique similar to that of Poborchii.[19,20] A monocular microscope was fitted with a central fiber optic light source with coaxial light pickup fibers (Ocean Optics). A polarizer was placed at the input of the microscope objective to obtain measurements of either transverse magnetic (TM) or transverse electric (TE) polarized light, and the sample was placed under the objective with an incidence angle of approximately 85 degrees, as shown in Fig. 15.2(b). The pickup fibers delivered the reflected light to an Ocean Optics spectrometer for analysis. A diffuse reflecting white ZnO surface was used as a standard to eliminate spectral effects from the source and the polarizer.

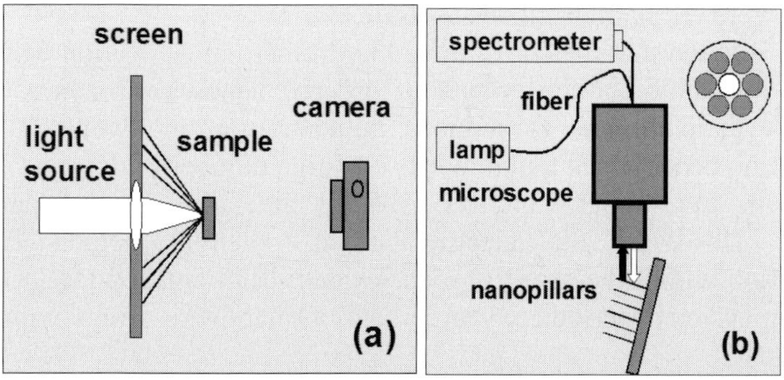

Fig. 15.2 Setup for optical diffraction (a) and in-plane reflection measurements (b). The upper right hand corner of (b) shows the cross section of the fiber with the central region for light delivery and the surroundings (gray region) for light collection.

15.3 Results and Discussion

Figure 15.3 shows the SEM images of a polymer template, the nanowire overgrowth on the template, and the nanowire arrays after the template was removed. An array of hexagonal patterns was created by electron beam lithography. After development, growth windows with sizes of about 5 μm were formed, as shown in Fig. 15.3(a). The lighter regions are the windows for ZnO nanowire growth, and the darker regions are PMMA. Figure 15.3(b) shows the nanowires grown on the templated substrate. The nanowires grew on both the growth windows and the PMMA. After liftoff, dense ZnO nanowire arrays were obtained in the growth windows while there are no nanowires left in the areas that were masked by PMMA.

It was demonstrated that large scale, spatially organized growth of ZnO nanowire clusters can be achieved by using a polymer template created by e-beam lithography. The size of the growth window shown in Fig. 15.3 is in the range of micrometers. However, the gap between the neighboring hexagonal patterns is only 200 nm. Clean liftoff of the wires suggests that those at the edges of the template are vertically aligned, and that the technique can be used for higher resolution work.

Fig. 15.3 SEM images of PMMA template (a), ZnO nanowires grown on the template (b), and the nanowires on substrate after liftoff of PMMA template (c). The light area in (a) is the substrate surface and the dark regions are PMMA template.

High quality arrays of ZnO were created, with individual pillars nucleated in the smaller PMMA apertures. Fig. 15.4 shows the SEM images of ZnO nanowires grown on patterned substrate with different circular aperture sizes. The growth conditions such as concentration, temperature, growth time and applied potential were the same for the samples displayed in Fig. 15.4. The ZnO nanowire arrays were confined to the holes as desired. After liftoff, the substrate surface that was previously covered by PMMA is free of nanowires. When the size of the opening was in the micrometer range, tens of ZnO nanowires were nucleated within the aperture (see, for example, Fig. 15.4(a)). As the pattern size was reduced, the number of nanowires decreased. As shown in Fig. 15.4(c), fewer than 4 nanowires grew on each window with size of 200 nm. If we assume the nucleation density to be constant as the pattern size changes, then the number of nanowires in each aperture would be proportional to the area of the opening. This assumption is confirmed by a plot shown in Fig. 15.4(d), which shows the number of nanowires in each aperture window as a function of the aperture area. The number of nanowires was counted directly from SEM images. The linear dependence of the number of nanowires on the aperture area indicates that the nucleation density of ZnO nanowires is independent of the size of growth window. The average nucleation density calculated from Fig. 15.4(d) is 8.5×10^9 / cm^2, which is consistent with the values previously observed in the electrochemical process[8].

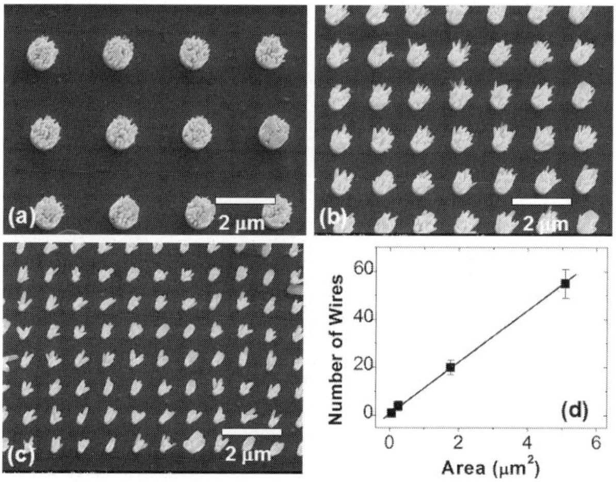

Fig. 15.4 SEM images of ZnO nanowires grown with different circular aperture sizes (a,b,c) and the number of nanowires on each pattern as a function of the area of the pattern (d).

It can be concluded from Fig. 15.4(d) that only one nanowire should grow in each aperture, under the conditions used, if the aperture diameter is less than 135 nm. This was demonstrated in our experiments as shown in Fig. 15.5. Figure 15.5(a) is an SEM image taken with a tilt angle of 30° which indicates the growth of individual ZnO nanowires on the templated substrate. The size of each growth window was around 100 nm with a period of 1000 nm. It can be seen that a single nanowire grows vertically on each site. Some of the nanowires were removed from the substrate during post growth processing, and are seen lying horizontally on the substrate. From this image, we can confirm the aspect ratio of the nanopillars -- they have a diameter of ~ 100 nm, and a length of approximately 1000 nm. Figure 15.5(b) shows the top view of the nanowire arrays and Fig. 15.5(c) is an image with a higher magnification. Each nanowire is precisely placed in the desired location and vertically aligned on the substrate with a hexagonal cross section, demonstrating single crystal ZnO growth along the (0001) direction.

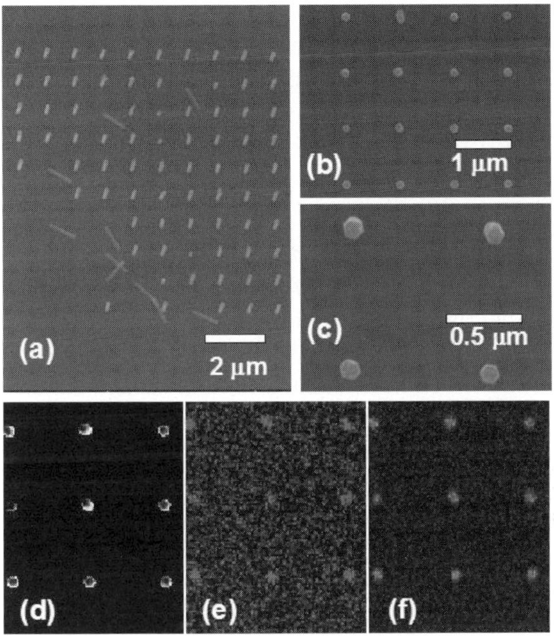

Fig. 15.5 (a) SEM images of individual ZnO nanopillars. Note the aspect ratio of the pillars that have been detached from the substrate; (b,c) top view of the nanowire arrays at higher magnification; (d,e,f) the corresponding image of electron emission (d), oxygen (e) and zinc signal (f) measured by EDX mapping.

The composition of the individual aligned nanowires was confirmed by EDX spectroscopy and mapping. Figure 15.5(d-f) shows the electron emission image and the corresponding composition maps of oxygen and Zn. The EDX mapping was achieved by scanning the electron beam over the measurement surface with detection energy fixed at the value of interest. The secondary electron emission image indicates the location of the nanowires. The compositional maps of Fig. 15.5(e) and 15.5(f) clearly indicate that the strong oxygen and zinc signals are from the nanowires. Further analysis of EDX spectra from the same sample suggests that the nanowires are ZnO. The relatively strong background in the oxygen map (Fig. 15.5(e)) is a result of oxidation of the substrate surface.

ZnO nanopillar arrays are potentially useful for photonic and optoelectronic applications that require precise control of the location of each pillar. The process used in this study is suitable for fabricating such two-dimension photonic structures and we confirmed the long-range order of our structures by light diffraction experiments. The experimental setup is illustrated in Fig. 15.2(a). A square lattice was used for this study. Fig. 15.6(a) shows an SEM image of the nanopillar array. Pillars with diameters of 140 nm, spacing of 0.98 μm and lengths up to 1 μm were fabricated by the template technique. Figure 15.6(b) is a Fourier transform (FFT) of the two-dimensional nanowire arrays, showing the expected symmetry of the diffraction pattern. The actual diffraction pattern of the nanowire arrays of Fig. 15.6(a) was measured, and is shown in Fig. 15.6(c). Note that the diffraction angles are large enough to exhibit distortion from a square pattern when displayed on a flat projection screen. The diffraction angle of the first order spots was calculated from the measured distance from the sample to the screen and the distance along the screen. The obtained diffraction angle of 34.2° is in good agreement with the value of 32.9° calculated directly from the diffraction equation ($d\sin\theta = m\lambda$). The small difference between the measured and calculated values indicates that the lattice constant of the two-dimensional nanowire arrays was well defined. The periodic organization of the nanowire arrays is also partially reflected by the small diffraction spots in the diffraction pattern.

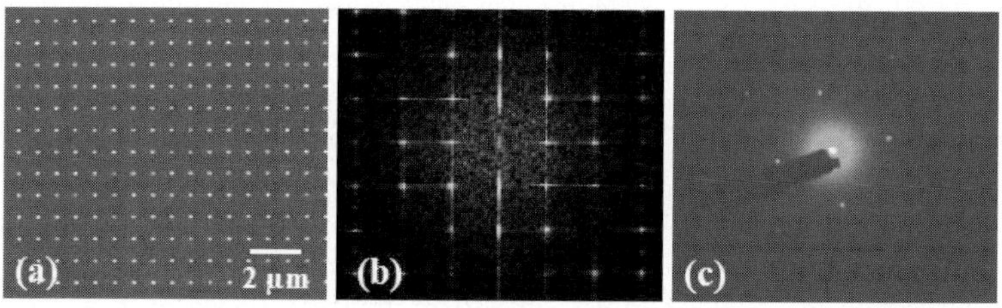

Fig. 15.6 (a) SEM image of a square lattice of a two-dimensional ZnO nanowire array; (b) corresponding Fourier transform pattern; and (c) light diffraction pattern using green light of 532 nm wavelength.

Measurement of the in-plane reflectance of an array is a straightforward method for establishing the potential of a structure for photonic band-gap applications. Fig. 15.7 shows the reflection spectrum along the Γ-X direction of the square array with lattice constant 480 nm under TM excitation. A reflection peak at 2.27 eV was observed, corresponding to a photonic band-gap. The band structure was calculated[21] and is shown in the inset of Fig. 15.7. Two TM band gaps with center energies of 1.1 and 2.15 eV were predicted along the Γ-X direction. The experimentally observed reflection peak originated from the second-order band of the two-dimensional crystals. The energy difference between theory and experiment is likely due to the combined effects of the uncertainty in the refractive index of ZnO, tilt effects and the finite height of the pillars. The first order band at 1.1 eV is outside the wavelength range of our measurements.

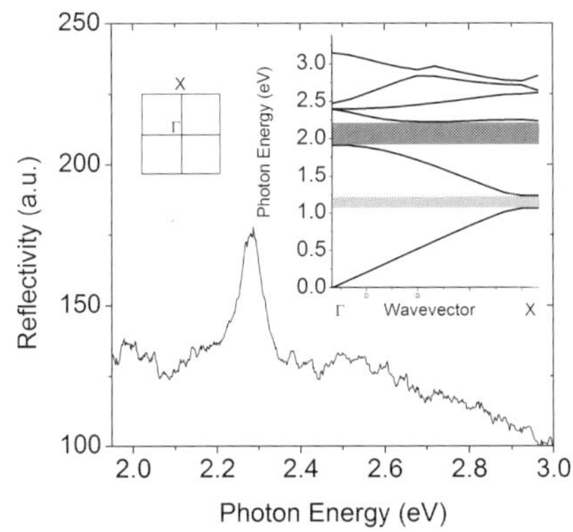

Fig. 15.7 Reflection spectrum of a square array of ZnO nanopillars in the Γ-X direction. The reflection from dense, randomly grown ZnO nanowires was used for background subtraction. The inset shows the calculated photonic band structure of a two-dimensional ZnO nanowire array with a lattice constant of 500 nm and nanowire radius of 130 nm. The gray regions represent the first and second order bands of the crystal.

For UV applications, a further reduction in the lattice constant of the arrays could be used to move the band to higher energies. Our calculations indicate that a ZnO nanowire square array with a lattice constant of 250 nm would exhibit a first-order TM photonic band-gap in the visible (2.3 eV) and a second order band-gap in the UV range (4.3 eV) along the Γ–X direction. A hexagonal photonic structure with a lattice constant of 220 nm exhibits a first-order band-gap of 2.4 eV[16]. Future experiments are needed to explore the density limits of our templating process.

ZnO is a promising candidate for photonic band gap materials for visible wavelengths. Its potential applications in photonic devices rely on the design of special nanostructured patterns and desirable defects within the photonic structure. Electron beam lithography allows the fabrication of arbitrary patterns for these applications, which constitutes a major advantage of using PMMA as template. We also demonstrated that arbitrary patterns of ZnO nanowire arrays can be created through the low temperature growth combined with the e-beam lithography-generated templates. As an example, two different patterns are shown in Fig. 15.8. The SEM image of Fig. 15.8(a) shows the growth of two rows of ZnO nanowires with one missing nanowire in the middle of each row. These sorts of structures could potentially find applications as templates for the construction of plasmonic resonators. In Fig. 15.8(c), a superlattice structure is shown, where each set of 7 nanopillars forms the basic unit cell of a hexagonal pattern which is repeated on the substrate surface to form a micro-periodic structure.

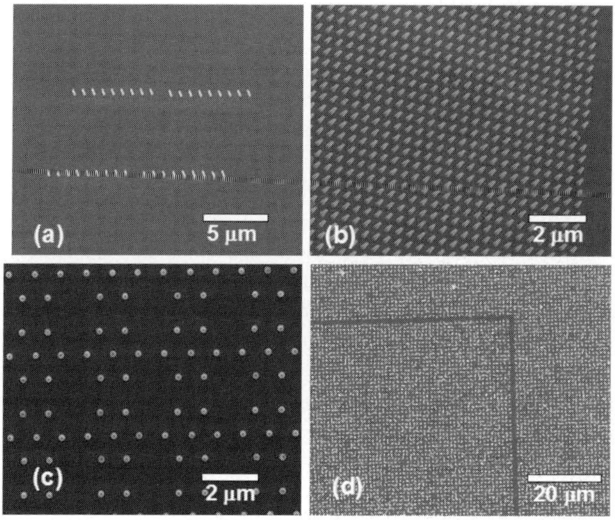

Fig. 15.8 SEM images of ordered ZnO nanopillar arrays with different layouts.

Since ZnO is a promising candidate for visible-wavelength photonic crystals[3], the design and fabrication of aperiodic nanostructures becomes a key step for corresponding applications[22]. As an example, we also fabricated a two-dimensional ZnO nanopillar array in which line defects were created. In a well-ordered two dimensional array, the nanowires were absent along half of a row and a column to form a transmission channel for light with a 90 degree bend. This channel can be used as a waveguide for light transmission within the two-dimensional photonic crystal[23]. This and related special structures are important for applications such as large scale all-optical circuits. Our two-dimensional ZnO nanowire photonic crystals have lattice constant varying from 300 nm to a few micrometers, which makes the two-dimensional nanowire arrays suitable for visible and near-IR photonics.

Although the throughput of e-beam lithography has been regarded as a major drawback for device fabrication, it is not a barrier for creating the photonic templates used in this study due to the very small effective writing area. Each writing window for individual nanowires growth is around 100 nm in diameter, therefore the total writing area is relatively small. The writing time was about 1 min for a 300 x 300 μm^2 template with feature size of 100 nm and period of 1 μm. In our processing, the aspect ratio of the arrays was controlled by the growth time and not the depth of the template. Thin layers of resist are compatible with the high density patterning required for photonic devices.

15.4 Conclusions

Highly ordered single crystal ZnO nanopillar arrays were successfully deposited on templated substrate by using an electrochemical growth process. The growth was performed in an aqueous solution, which has numerous advantages such as moderate temperature (90°C), scalability, and compatibility with polymer template materials. Without the template, ZnO nanowires grew randomly on the substrate surface, with a range of alignment angles. The use of the template defines the growth location of nanowires and promotes growth normal to the substrate surface. We demonstrated that for small apertures, spatially organized ZnO nanopillar arrays with arbitrary pattern structures and/or intentional defects can be grown using a PMMA template created by electron beam lithography. This process can be used for fabricating large-scale and well ordered two-dimensional ZnO photonic materials. We also demonstrated that photonic crystals with a lattice constant of 480 nm exhibited a second harmonic band gap in the visible spectrum for TM polarized light; consistent with theoretical calculations. This fabrication approach is compatible with current microfabrication processes and has the potential for scalability and for fabricating a variety of ZnO photonic devices.

Acknowledgement

The author would like to thank Dr. Ursula Gibson for her calculations of photonic band structure and helpful discussions.

References

1. Y. G. Roh, S. Yoon, S. Kim, H. Jeon, S. H. Han, Q. H. Park and I. Park, Appl. Phys. Lett. 83, 231 (2003)
2. T. Tada, V. V. Poborchii and T. Kanayama, Microelectr. Eng. 63, 259 (2002)
3. C. S. Kee, D.K. Ko and J. Lee, J. Phys. D: Appl. Phys. 38, 385 (2005)
4. M. Huang, Y. Wu, H. Feick, N. Tran, E. Weber and P. D. Yang, Adv. Mater. 13, 113 (2001)
5. Z. L. Wang, J. Phys.: Condens. Matter. 16, R829 (2005)
6. J. B. Cui, C. P. Daghlian, U. J. Gibson, R. Pusche, P. Geithner and L. Ley, J. Appl. Phys. 97, Art. No. 044315 (2005)
7. L. E. Greene, M. Law, J. Goldberger, F. Kim, J.C. Johnson, Y. Zhang, R. J. Saykally and P. D. Yang, Angew. Chem. Int. Ed. 42, 3031 (2003)
8. J. B. Cui and U. Gibson, J. Phys. Chem. B 109, 22074 (2005)
9. L. Vayssieres, Adv. Mater. 15, 464 (2003)
10. P. D. Yang, H. Q. Yan, S. Mao, R. Russo, J. Johnson, R. Saykally, N. Morris, J. Pham, R. R. He and H. J. Choi, Adv. Funct. Mater. 12, 323 (2002)
11. E. C. Greyson, Y. Babayan and T. W. Odom, Adv. Mater. 16, 1348 (2004)
12. H. T. Ng, J. Han, T. Yamada, P. Nguyen, Y. P. Chen and M. Meyyappan, Nano Lett. 4, 1247 (2004)
13. X. D. Wang, C. J. Summers and Z. L. Wang, Nano Lett. 4, 423 (2004)
14. H. J. Fan, B. Fuhrmann, R. Scholz, F. Syrowatka, A. Dadgar, A. Krost and M. Zacharias, J. Cryst. Growth 287, 34 (2006)

15. J. W. P. Hsu, Z.R. Tian, N. C. Simmons, C. M. Matzke, J. A. Voigt and J. Liu, Nano Lett. 5, 83 (2005)
16. M. Matsuu, S. Shimada, K. Masuya, S. Hirano and M. Kuwabara, Adv. Mater. 18, 1617 (2006)
17. Y. J. Kim, C. H. Lee, Y. J. Hong, G. C. Yi, S. S. Kim and H. Cheong, Appl. Phys. Lett. 89, 163128 (2006)
18. J. B. Cui and U. J. Gibson, Nanotechnology 18, 155302 (2007)
19. V. V. Poborchii, T. Tada and T. Kanayama, Appl. Phys. Lett. 75, 3276 (1999)
20. V. V. Poborchii, T. Tada and T. Kanayama, J. Appl. Phys. 91, 3299 (2002)
21. S. G. Johnson, http://jdj.mit.edu/wiki/index.php/MIT_Photonic_Bands
22. A. Mekis, J. C. Chen, I. Kurland, S. H. Fan, P. R. Villeneuve and J. D. Joannopoulos, Phys. Rev. Lett. 77, 3787 (1996)
23. S. Y. Lin, E. Chow, V. Hietala, P. R. Villeneuve and J. D. Joannopoulos, Science 282, 274 (1998)

16

VISIBLE LIGHT EMISSION FROM INNATE SILICON NANOPARTICLES IN SILICON-COMPOUND FILMS GROWN AT LOW TEMPERATURES

Cao Zexian

This chapter describes Plasma Enhanced Chemical Vapour Deposition (PECVD) growth of silicon nanoparticles embedded in silicon oxide, nitride and carbide films. The influence of plasma forming discharge parameters on the size and density of the embedded silicon nanoparticles is described. Light emission across the whole of the visible light range was observed in both the Si-in-SiO_x and Si-in-SiN_x structures. Of the two, the latter was found to be the superior system because of its high external quantum efficiency and short radiative decay time.

16.1 Introduction

Silicon has served as the "work horse" material for the microelectronics industry where electrons are employed to transmit information. The industry has already perceived its end of development with regard to further raising the density of information storage and the speed of information processing. Optoelectronics and photonics are technologies that partially or entirely use light for information manipulation. By replacing electronics with optoelectronics or photonics we can expect devices to run at incredible speeds. Other advantages of using light instead of electrons include reduced power consumption, enhanced reliability, higher volume efficiency, etc. Silicon and other Si-based materials, if they can emit strong light and allow relatively unsophisticated device fabrication, will be the first choice for optoelectronics and photonics. This is so because with its use decades-old silicon manufacturing expertise could still be exploited[1], particularly when integration with electronics is of concern. Enormous effort has been devoted to the fabrication of silicon nanostructures that are expected to emit light across a wide, tunable spectral range, and with respectable efficiency[2-4]. Among the various possibilities, light emission from silicon nanoparticles embedded in confining films deserves particular attention. In this chapter we shall present a brief review of a low-temperature fabrication procedure involving conventional plasma-enhanced chemical vapor deposition (PECVD) method, by which high-density innate silicon nanoparticles are grown along with a silicon compound (oxide, nitride or carbide) film, thus making a composite nanostructure which has demonstrated excellent photoluminescence (PL) features. These include tunability across the whole of the visible light range[5-13] and luminescence decay-time of a few nanoseconds[10,13]. After a brief introduction, we'll first discuss the formation of the silicon particles in silane plasma which is the key process for the deposition of the silicon nanoparticles in its compound films and then present our experimental results on the light-emitting Si-in-SiO_x, Si-in-SiN_x and Si-in-SiC composite structures.

Bulk silicon is an indirect bandgap (E_g ~1.12 eV) semiconductor which emits light with a negligibly small efficiency. By reducing the size of the silicon agglomerates to a size of a few nanometers, the electrons now having a wavelength larger than the typical dimension of the materials structure (the Bohr radius of excitons in silicon is ~5.0 nm) sense the presence of the geometrical boundaries i.e. the quantum confinement effects (QCE) become obvious. Strong photoluminescence is anticipated in silicon nanostructures[14-16], as was first observed in porous silicon in 1990[17]. However, due to mechanical fragility and physical/chemical instability, porous silicon is not a promising material for silicon-based light-emitting devices (LEDs). On the other hand, silicon particles embedded in a host material, generally the silicon oxide, nitride or carbide matrix feature both effective confinement and attain a robust structural stability that is very desirable for the fabrication of devices. A variety of different approaches have been developed for the preparation of such Si-in-silicon compound systems. Taking Si-in-SiO_x composite films for instance, the preparation methods include Si-ion implantation into silicon oxides, sputtering, reactive evaporation or CVD growth of Si-rich oxides, as well as post-annealing of SiO/SiO_2 multilayers etc[18]. These methods often involve high temperatures that approach 1100°C to obtain nanocrystalline silicon particles by post-annealing treatments. Such a high temperature process is clearly prohibitive from the technical viewpoint of device integration since it causes severe degradations effects in existing devices by promoting dopant diffusion. A modus operandi for the preparation of silicon-based light-emitting nanostructures that employs only low temperatures is obviously preferable. Fortunately, PECVD provides a convenient low-temperature route for this purpose.

16.2 Formation of Silicon Particles in Silane Plasma

It has long been noticed that in an RF plasma formed from gas mixtures containing silane, large numbers of silicon clusters are observed to form in the discharge even at a low working pressure (13 Pa) and low power density (about 0.1 W/cm^2)[19,20]. Small silicon clusters begin to nucleate at the plasma/sheath boundary near the electrodes and principally grow in the same region. The size of the clusters near the substrate on the grounded electrode is generally less than 15 nm and the density can be as high as 10^{11} cm^{-3}. Some silicon particles may be incorporated into the growing films since most of them are generally neutral, though the sticking coefficient at the surface is quite low (<5%). This clearly leads to a serious contamination problem for the deposition of highly homogeneous a-Si:H films and this, in turn, causes a reduction in the yield of devices. It is, however, advantageous for the low-temperature deposition of composite films wherein innate silicon particles get embedded in an insulating barrier layer of silicon compounds.

In order to obtain strong photoluminescence from the embedded silicon particles in the visible spectral range, effective control of particle size, passivation conditions and particle density is essential. This can partially be realized by adjusting the plasma conditions through changing the total pressure, the power density and/or the fraction of silane in the precursor. A large amount of work has been done to raise the particle density to enhance emission strength since whatever the mechanism may be the light emission is proportional to the number of emission centers. As to the particle size, it is conceivable that the introduction of oxidizing ion species, such as that of oxygen, into the plasma can reduce the size of silicon particles to below 5.0 nm. This then provides an effective route for the tailoring of particle sizes to a suitable value for light emission centered at desired wavelengths. Of course, changing the size of silicon particles by varying the plasma conditions will also influence the density of the silicon particles coming to the growing surface, yet an independent change of particle density from the particle size can be realized in a narrow region by fine tuning the substrate temperature.

By taking advantage of silicon particle formation in the plasma sheath, we have succeeded in growing Si-in-silicon compound composite structures onto cold substrates. High-efficiency PL emissions at desired wavelengths have been achieved through increasing particle density and tailoring particle sizes. In some cases a heat-treatment was applied to enhance the overall quality of the film or to improve the passivation of silicon particles but the temperature involved was generally below 500°C. Later we will show that this post-treatment at elevated temperatures is completely unnecessary for the Si-in-SiN_x and Si-in-SiC systems[10-13].

16.3 Experimental Details

Si-in-silicon compound films were deposited onto single-crystal Si (001) wafers and fused quartz substrates using a custom-designed capacitance-coupled PECVD setup. Precursors were a gas mixture of high-purity SiH_4 and H_2 with the addition of O_2 or N_2O for the oxide, of N_2 for the nitride, and of CH_4 for the carbide system, respectively. The fraction of SiH_4 in the precursor served as the controlling parameter, while the total flow rate was generally being maintained constant. At a working pressure of ~ 1 x 10^2 Pa and an RF power density as low as 700 mW/cm^2, the resulting growth rate was estimated to be a few nm / min; depending on the species of the reactive gases. No intentional substrate heating was applied; the substrate temperature, altered exclusively by the plasma irradiation was restricted below 60°C. In order to maintain such a low substrate temperature without involving any cooling, we often adopted a cyclic growth mode i.e. after every hour or so when the substrate temperature reached 50°C, the power supply would be turned off to let the system cool down to room temperature. This method also helps improve the film transparency and thus the efficiency of light emission. The final film thickness generally spanned a range of 0.5 – 1.5 μm.

The local atomic environments and bonding configurations in the deposits were characterized by infrared absorption measurements performed on a Perkin-Elmer 2000 Fourier-transform infrared system and by inspecting the Si 2p line excited with Mg Kα radiation on an X-ray photoelectron spectrometer (ESCALab Mark-II). Formation of silicon nanocrystals in the films was also verified by micro-Raman scattering using a confocal laser Raman spectrometer (JY-6400). The size and spatial distribution of the innate silicon nanoparticles were investigated from the bright-field images obtained with a Tecnai-20 transmission electron microscope (TEM) for which a proper defocus value had to be chosen to accomplish an enhanced contrast. Size determination suffers from large uncertainties due to the amorphous nature of the silicon particles and the less pronounced particle/matrix contrast. In the case of very high particle density, agglomeration of a few particles into a weakly-connected cluster also occurs.

As-deposited films were first examined under an ultraviolet lamp to confirm the presence of PL. Room-temperature PL measurements were carried out on an FLS 920 (Edinburgh Instruments Ltd.) fluorescence spectrometer with a xenon lamp as the excitation source. Time-resolved analysis was performed on a spectrophotometer (Horiba NAES-1100) with excitation at 375 nm from a hydrogen lamp (chopper frequency 10 kHz, pulse width 2.0 ns) to determine the decay times at different emission wavelengths registered within less than 1.0 nm. The temporal resolution of the SSU-112 photomultiplier employed was 0.1 ns. The external quantum efficiency was determined by spectrophotometric measurement with the aid of an integrating sphere which collected both the reflected and emitted lights without wavelength discrimination. Other parameters or methods associated with the preparation or characterization of a particular sample will be specified where needed.

16.4 Light Emission from Innate Silicon Particles

Highly transparent Si-in-SiO_x, Si-in-SiN_x and Si-in-SiC films containing a large density of innate silicon nanoparticles were successfully prepared by using our low-temperature procedure. The particles can be easily tailored to various small sizes so as to generate strong PL across the whole visible light range. We have made special efforts to explore the smallest size of the silicon particles attainable in a specific matrix in order to determine the shortest wavelength for the quantum confined PL. The particle density generally amounts to 10^{12} – 10^{13} cm^{-2} (from TEM micrographs). This results in very strong PL which essentially facilitates the measurement of the quantum efficiency, decay time and other important parameters. Detailed results are presented below in the order of Si-in-SiO_x, Si-in-SiN_x and Si-in-SiC.

16.4.1 Si-in-SiO_x

The Si-in-SiO_x system immediately attracted the attention of researchers after the discovery of light emission from porous silicon. In the beginning, the effort had been focused on composite nanostructures containing silicon nanocrystals dispersed in SiO_2[2,21,22]. The requirement on the nanocrystallinity of silicon particles is believed to enable both increase in the radiative electron–hole recombination rate and reduce the non-radiative pathways, thus increasing the emission efficiency. Here we will show that an effective confinement does not alter the crystallinity of silicon particles. These silicon nanocrystals can be grown by a PECVD process onto substrates at room temperature. An enhancement up to two orders of magnitude in the PL efficiency can be achieved in films grown on cold substrates, followed by a subsequent rapid thermal annealing (RTA) at 500°C for two minutes in a N_2/Ar ambient atmosphere.

Si-rich oxide film is a precursor for the preparation of silicon nanocrystals in stoichiometric SiO_2. Si-rich oxide films that contain some innate silicon particles are usually grown by using CVD at a substrate temperature between 250 – 350°C. This is in conformity with requirements for fabricating device-quality hydrogenated amorphous or microcrystalline silicon films. However, there is only a limited degree of phase separation between the silicon particles and the oxide. The oxide phase is sub-stoichiometric SiO_x with the value of x increasing along with the distance from the silicon core[23]. This results in a limited density of silicon particles and poor carrier-confinement. Consequently, the room-temperature PL from the as-deposited films is several orders of magnitude less intense than that from silicon nanocrystals in SiO_2.

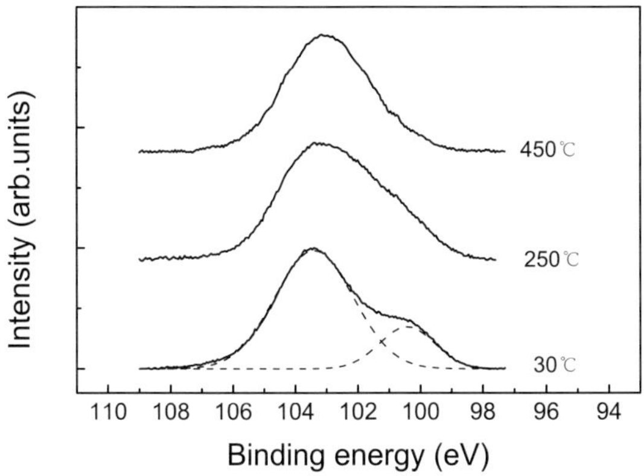

Figure 16.1 Evolution of the Si 2p line for three silicon-rich oxide films grown at different substrate temperature.

We noticed that even though the silicon particles were formed in the plasma sheath, the substrate temperature was still a key factor that influenced the structure of the deposits. Enhanced diffusion at the growing surface at elevated temperatures may pre-exclude the presence of distinct silicon particles. A very low temperature is more favorable for the survival of silicon particles in a silicon compound matrix. Fig. 16.1 illustrates the evolution of the Si 2p line from XPS measurements on three si-rich silicon oxide films prepared with, otherwise, the same processing parameters. These had roughly the same composition (averaged atomic ratio γ = [O]/[Si] ~1.5). The Si 2p line for the film deposited at 450°C displays only one nearly symmetric peak, which can be attributed to the Si-Si_xO_{4-x} (x=0, 1, 2, 3, 4) tetrahedral units of rather homogeneous distribution, as predicted by the random-bonding model. For the sample deposited at 250°C, it became somewhat asymmetric and broadened at the low binding-energy side due to the increased concentration of the Si-Si_4 structure. In contrast, the film grown at 30°C showed a distinct two-peak profile which can be well fitted with two Gaussian profiles centered at 100.3 eV and 103.4 eV, respectively. This strongly suggests a film composition of well-separated Si and SiO_2 phases. The slightly enlarged binding energy at 100.3 eV for the a-Si particles is mainly caused by the charging effect in this semi-insulating composite film.

The phase separation in the deposits can be directly verified by the bright-field TEM images. In Fig. 16.2, the silicon nanoparticles (dark

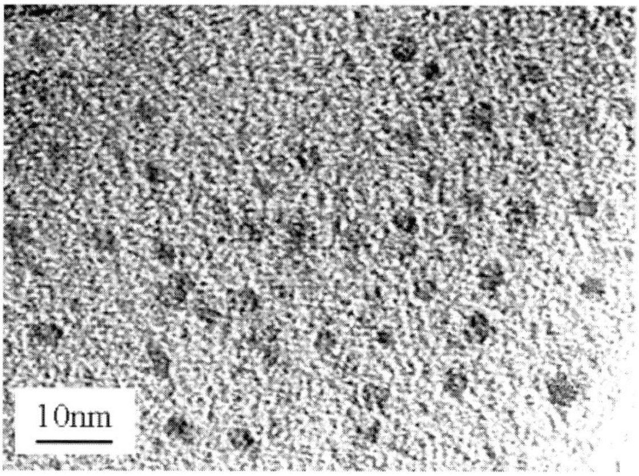

Figure 16.2 TEM image of a silicon-rich oxide films with an overall atomic ratio $\gamma \sim 1.5$ deposited at 30°C.

spots), are essentially amorphous, have an averaged size of about 3.0 nm and the number density is obviously over $10^{12}/cm^2$ (from the micrograph). From the PL measurements (Fig. 16.3a) we can see that the film deposited at 30°C, characterized as Si-in-SiO_x, emits at an intensity over one order of magnitude larger than that grown at 250°C; both as deposited and after the RTA treatment. In addition to the redshift of the PL spectrum, post-annealing also results in a considerably enhanced PL efficiency. For the sample deposited at 30°C, the integrated intensity of the PL spectrum increases by a factor of 8. This enhancement of PL by post-annealing cannot be emic rearrangements in silicon oxide that can eventually leaxplained by the additional precipitation of new silicon nanoparticles, since a post-annealing at 500°C for two minutes does not cause any noticeable atod to the formation or growth of silicon particles. Instead, we noticed that the RTA treatment at 500°C can effectively release hydrogen atoms incorporated in the deposits, thus modifying the configuration of some local hydrogen-bonded structures. This occurs particularly at the particle/matrix interfaces, as confirmed by infrared absorption measurements. This reasoning is also supported by the red shift of PL, since the optical band gap of silicon particles becomes smaller when hydrogen is released such that the hydrogen passivation of the silicon particles is partially replaced by oxygen-passivation[24].

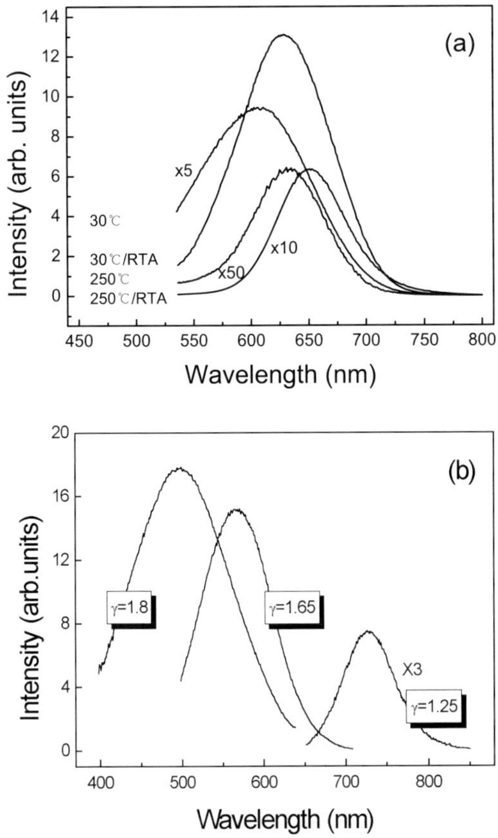

Figure 16.3 (a) PL spectra of the as-deposited and annealed Si-rich oxide films referred to in Fig. 16.2; (b) PL spectra of the annealed silicon-rich oxide films deposited at 30°C but of different compositions.

The size of the silicon particles in films can be conveniently reduced by increasing the proportion of the oxygen in plasma and thus the proportion of oxygen in deposits. We see that the peak wavelength of the PL emission has been shifted to below 500 nm and the external quantum efficiency can be easily made comparable to that for silicon nanocrystals (Fig. 16.3).

From the standpoint of device fabrication even the RTA at 500°C is undesirable. How simple would it be if the silicon nanocrystals could be grown in the as-deposited film in such a manner that the characteristics for both the particles and the compound matrix were simultaneously determined, once and for all. We have found that SiO_x films containing silicon nanocrystals could be synthesized by PECVD at 300°C or 400°C[23], but the sizes of the crystals are too large (over 6.0 nm) for quantum confined light emission.

By largely increasing the proportion of hydrogen ions in plasma (the summed flow rate for SiH_4 and O_2 being maintained at 2.0% of the total), silicon oxide films containing high-density innate silicon nanocrystals can be grown. It is thought that the large silicon particles formed in the plasma may include a well crystallized core. Effective chemical etching by hydrogen ions help preferentially remove the amorphous part. The hydrogen ions lead to the occurrence of elemental silicon even in deposits grown with an oxygen fraction r, defined as the flow rate of oxygen with regard to the summed flow rate for oxygen and silane, as high as r = 0.25.

The nanocrystallinity of the tiny silicon particles in the as-deposited films was confirmed by micro-Raman scattering measurements. In Fig. 16.4 the Raman spectra of samples prepared with different oxygen fractions are plotted. The sharp peak around 510 cm^{-1}, arising from the first-order Raman scattering of silicon phonons, confirms the presence of silicon particles in the form of nanocrystals. The significant red shift (i.e., softening of the phonon mode) and broadening of the Raman peak with respect to that for silicon bulk crystal are typical size

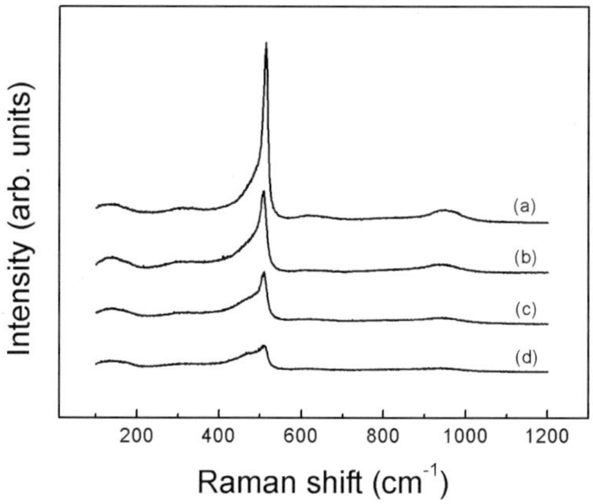

Figure 16.4 Micro-Raman spectra measured on samples grown with different oxygen fractions in precursor: (a) $r = 0.10$; (b) $r = 0.15$; (c) $r = 0.20$; and (d) $r = 0.25$. The sharp peak at around 510 cm^{-1} is characteristic of silicon nanocrystals.

effects in nanocrystals. In the more heavily oxidized samples, the line width of the Raman peak increases while its intensity quickly diminishes. This suggests a gradual reduction in the size of silicon nanocrystals, as was also verified by TEM micrographs. By taking into account the effect of both phonon confinement and elastic strain upon the Raman scattering of small crystals, the crystallite size was estimated to decrease, from curves (a) to (d), from 4.4 nm to 2.3 nm.

Since the silicon nanocrystals in these deposits are smaller than 5.0 nm, intense red PL was measured at room temperature. Remarkably, the peaks of the PL profiles all fall within a narrow range between 768 nm and 743 nm. This means that the PL is roughly pinned at a photon energy of 1.61 - 1.67 eV. This is quite difficult to understand from the viewpoint of quantum confined emission. Wolkin and coworkers once presented a detailed study on the possible PL mechanisms for silicon nanocrystals passivated with oxygen by calculating the electronic structures of such giant molecules[25]. They found that a stabilized electronic state or even a

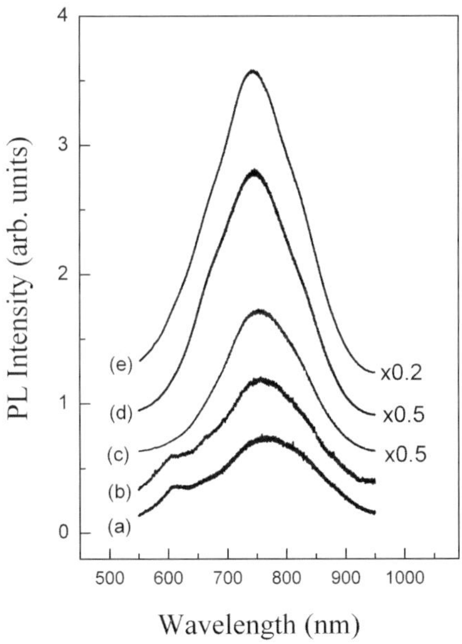

Figure 16.5 PL spectra from as-grown nc-Si-in-SiO_x samples prepared with varied oxygen fraction in precursor: (a) r = 0.10; (b) r = 0.12; (c) r = 0.15; (d) r = 0.20; (e) r = 0.25.

trapped exciton may be formed on the $Si = O$ double bonds. In addition, they found that for particles smaller than 3.0 nm, the PL photon energy increases only very little (on the contrary, in amorphous particles, optical transition other than from the trapped exciton states dominates due to the lack of long-range order within the particles). The photon energy measured by us (1.61 - 1.67 eV) falls on the theoretical curve for the lowest transition in the presence of $Si = O$ double bonds, but smaller than the upper limit of the emission energy of 2.1 eV. The significant enhancement in the PL efficiency for the silicon nanocrystal samples prepared with more oxygen in precursor (Fig. 16.5d-16.5e) originates mainly from the substitution of Si-H passivation by $Si = O$ double bonds.

16.4.2 Si-in-SiN$_x$

Although high-efficiency light emission at room temperature has been realized in various nanostructures based on silicon oxide several problems remain to be resolved for the realization of silicon-based LEDs. The external quantum efficiency needs be raised to an acceptable level, the emission profiles should be further improved for better color purity and shifted towards shorter wavelengths and the decay time should be brought down to within one nanosecond or even shorter.

One particular problem concerns carrier injection for electroluminescence. Carrier injection is performed through the tunneling mechanism. For silicon particles embedded in SiO_2, which has a bandgap of about 8.5 eV, the operating voltage of LEDs would be unacceptably high. Clearly, if silicon nanoparticles that give off strong light emissions can be prepared in matrices of smaller bandgap material then the technology of silicon LEDs could be developed for commercial exploitation. One immediate substitute for SiO_2 is silicon nitride, having a narrower bandgap of ~ 5.3eV, which has been widely used in electronics. Providing that Si-in-SiN_x films of comparable light-emitting ability[26] can be prepared by the low temperature procedure we presented above, the carrier injection problem for electroluminescence might be solved. Moreover, replacing the oxide matrix with nitride also could help clarify the controversial mechanism underlying light emission from silicon nanostructures[27,28]. Of course, one prerequisite for any further discussion is the realization of high-efficiency PL from the Si-in SiN_x system.

First, we present the result of investigations on a set of Si-in-SiN_x films grown by varying the nitrogen-to-silane flow rate ratio r between 1.0 and 5.5. The atomic ratio [N]/[Si] for the deposits increases roughly from 0.65 to 1.05. By taking into account the dispersed silicon particles, the stoichiometry for the SiN_x matrix was found to change only slightly, with the x value lying in the range of 1.0 to 1.10. With an increasing r value, the vibration frequency of the Si–N and Si-H stretching modes in the IR absorption spectra shift continuously towards larger wavenumbers, following a trend that would be expected from SiN_x samples containing more nitrogen.

The presence of silicon particles in the deposits again can be confirmed by Raman spectroscopy as for the Si-in-SiO_x system. This reveals an enhanced short-range and medium-range disorder of the Si-Si_4 network, as a consequence of reduced particle size in films with less silicon content. We see that silicon nanoparticles are isometrically distributed throughout the film (Fig. 16.6). Depending on the fraction of silane in the precursor and hence the silicon concentration in the deposit, the particles range in size from 1.5 nm to 4.5 nm. This corresponds to a number density between 1 and 4 x 10^{12} cm^{-2}. These particles are essentially amorphous as the electron diffraction pattern displays only one diffuse ring.

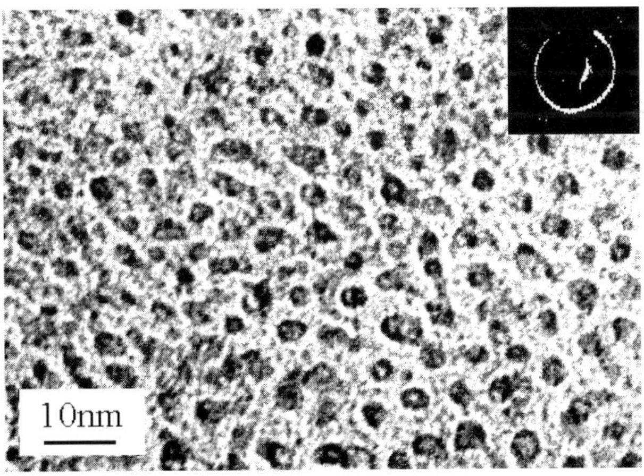

Figure 16.6 TEM image of a Si-in-SiN_x film grown with r = 2.0. Silicon particles (dark spots) of roughly 3 nm in diameter are clearly discernible. Particle number density is 3.4 x 10^{12} cm^{-2}.

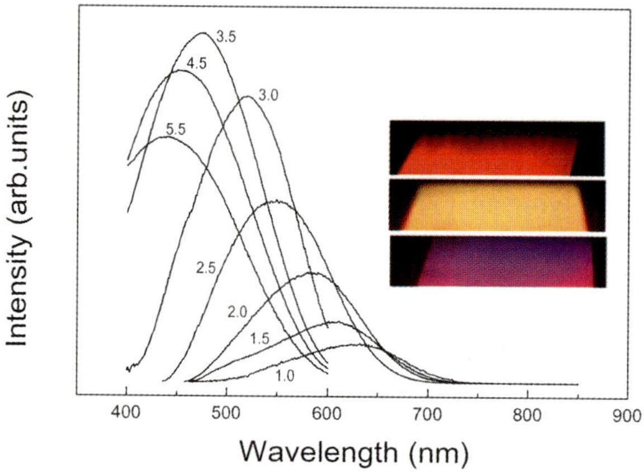

Figure 16.7 PL spectra from the annealed Si-in-SiN_x samples. Inset presents photographs of the light-emitting samples that were grown on glass with r values of 1.5, 3.5 and 5.5, respectively.

After RTA treatment, the samples exhibited strong PL in the whole of the visible spectrum (Fig. 16.7). The very bright red, green and blue emissions were perceptible to the unaided eye when illuminated with a 6-Watt UV-lamp emitting at around 365 nm. The external quantum efficiencies of PL for these annealed Si-in-SiN_x samples were estimated to be in the range of 0.5 - 5% from the red to blue end of the visible spectrum. Some persuasive observations support the view that PL from this Si-in-SiN_x system originates in the silicon particles rather than the nitride matrix. First, both PL intensity and peak position depend strongly on the density and size of silicon particles. The blue shift of the PL peak is unambiguously correlated to a reduction in the particle size. Second, PL from these samples is rather steadfast. Neither prolonged continuous illumination nor long term aging in air caused any noticeable degradation in intensity or modification to the PL profile.

The superior emission capability of the Si-in-SiN_x system over the Si-in-SiO_x system can be easily seen from Fig 16.8 where the PL peak intensity is plotted versus peak position for the two systems. In both cases, qualitatively speaking, the PL at shorter wavelengths corresponds to smaller silicon particles. Clearly, the PL from Si-in-SiN_x films has a higher efficiency than that from Si-in-SiO_x films, especially in the spectral range that corresponds to blue and green light. A factor of 3.4

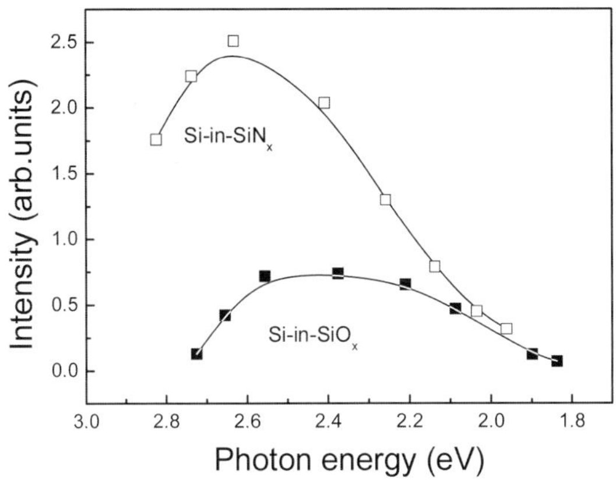

Figure 16.8 Comparison of PL peak intensity versus peak position between Si-in-SiN$_x$ and Si-in-SiO$_x$ thin films prepared with similar processing parameters. The lines are drawn to guide the eye.

enhancement was calculated at the photon energy of 2.6 eV. This is a reasonable consequence of the change in embedding matrix since the N ions are less reactive than O ions; favoring the formation of stable Si-SiN$_x$ interfaces. The density of silicon particles in SiN$_x$ is thus larger than that in SiO$_x$, especially when the particles are smaller. Besides, we also noticed that silicon particles of the same size in SiN$_x$ give rise to PL at shorter wavelengths. The detailed mechanism is not understood at the current stage of research.

The PL from silicon nanoparticles in SiN$_x$ and SiO$_x$ matrices show rather dissimilar annealing behavior. The peak position for Si-in-SiN$_x$ system progressively blue shifts by post-annealing below 600°C, while in the case of Si-in-SiO$_x$ it already red shifts at 360°C. Nevertheless, we could not attribute the PL peak shift induced by annealing to the size effect since no appreciable alteration in the particle size was evidenced from either TEM or Raman characterizations. This difference comes about from the modification of local hydrogen-bonded structures in these two material systems[5, 8].

The exciting results of PL measurements on the Si-in-SiN$_x$ system encouraged us to further explore this composite structure. Clearly, in Si-in-SiN$_x$ the PL blue shifts continuously with the decreasing particle size and it can be pushed to even shorter wavelengths than in the Si-in-SiO$_x$

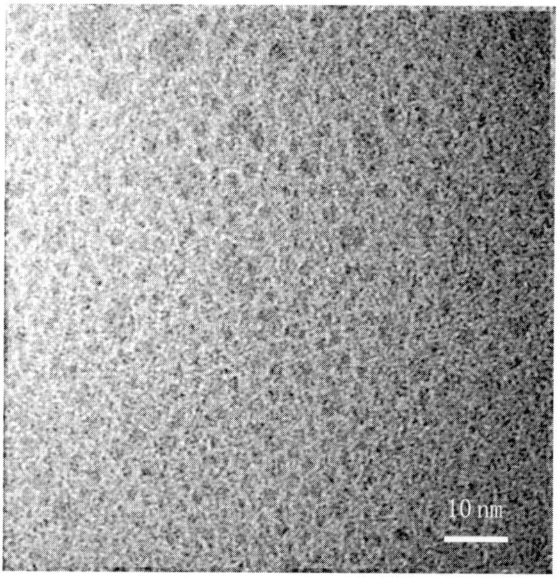

Figure 16.9 TEM micrograph of an as-deposited Si-in-SiN_x sample with [N]/[Si] ~ 0.77. The particle size falls within 2.2 nm - 2.6 nm range with only rare exceptions. The few large particles (> 5 nm) are due to the conglomeration of smaller ones.

system. Then what is the shortest PL wavelength attainable in the Si-in-SiN_x system? To answer this question, we further raised the nitrogen fraction in the precursor. By $r = 16$, a sample with the overall atomic ratio [N]/[Si] ~ 0.77 was prepared in which the size of the amorphous silicon particles was within the range of 2.2 nm to 2.6 nm and the particle density measured ~ 1.4 x 10^{13} cm^{-2} (Fig. 16.9). In this film, the PL was centered at λ_0 = 440 nm (2.82 eV). A further reduction in particle size can be realized at the cost of particle density, since it is difficult for the critically small particles to survive the oxidative matrix. In a sample prepared with gas flow rates, SiH_4:N_2:H_2 = 8.0:8.0:10 (here SiH_4 was previously diluted with hydrogen to 2.0 vol. %), the typical particle size was reduced to around 2.0 nm, but with a smaller density of 1.07 x 10^{13} cm^{-2} (Fig. 16.10). Due to the amorphous nature of the sample and the extremely high particle density, clusters of a few weakly connected particles formed. It is difficult to perform an exact statistical analysis of the particle size distribution but among the ~ 210 particles in Fig. 16.10, about one third are distinctly isolated ones with a dimension around

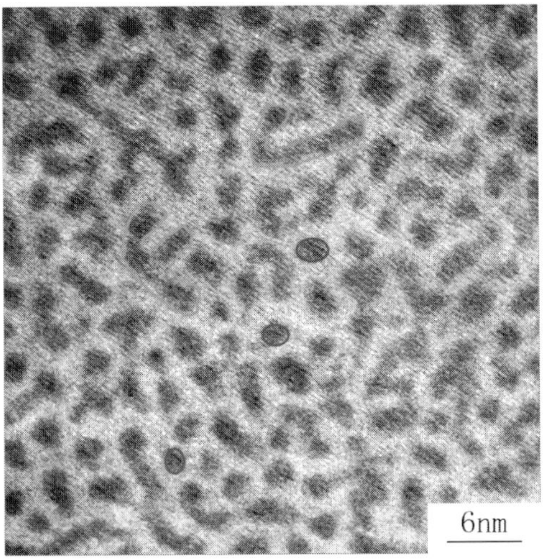

Figure 16.10 TEM micrograph of a Si-in-SiN$_x$ sample. Typical size of the silicon particles (circled) is ~ 1.8 nm but some weakly connected small clusters are also identifiable.

1.8 nm. In fact, while showing silicon particles of the smallest size, Fig. 16.10 is also a micrograph of the best quality we have ever obtained with the amorphous Si-in-SiN$_x$ and Si-in-SiO$_x$ films. The sharp particle/matrix boundary results from the cyclic growth mode; it has an essential to play in raising the efficiency of light emission.

With these extremely small silicon particles, the peak position of PL emission from Si-in-SiN$_x$ system was shifted down to 428 nm. We failed to shift the PL to even shorter wavelengths, with comparable intensity, by further reducing the fraction of silane. This was because silicon particles at such a high density, due to their small size, find themselves in a critical situation such that they either tend to get engulfed by the matrix compound or agglomerate to form bigger particles. A remarkable feature of the PL spectra in Fig. 16.11 is their enormous width. Even for curve (a) having the largest intensity, the full width at half maximum is 70 nm, indicating a very serious effect arising out of size fluctuations. The external quantum efficiency for the sample referred to in curve (a) was determined to be over 3.0%. For details of measurement and data processing see Ref. 10. This is consistent with the large internal quantum efficiency (over 10%) determined from the temperature-dependent PL measurements.

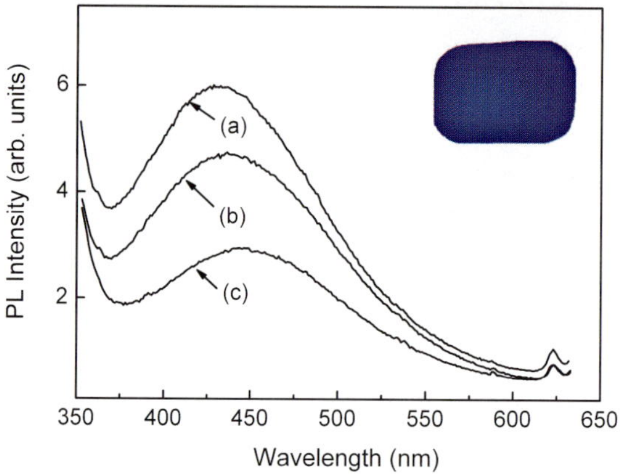

Figure 16.11 Room temperature PL spectra measured in three Si-in-SiN_x films grown with the gas flow rates SiH_4:N_4:H_2, from (a) to (c), of 8.0:8.0:10, 13:8.0:10 and 18:8.0:10, respectively. Inset (bright blue) displays a photo of the sample referred to in curve (a) under the irradiation from a 6 W ultraviolet lamp.

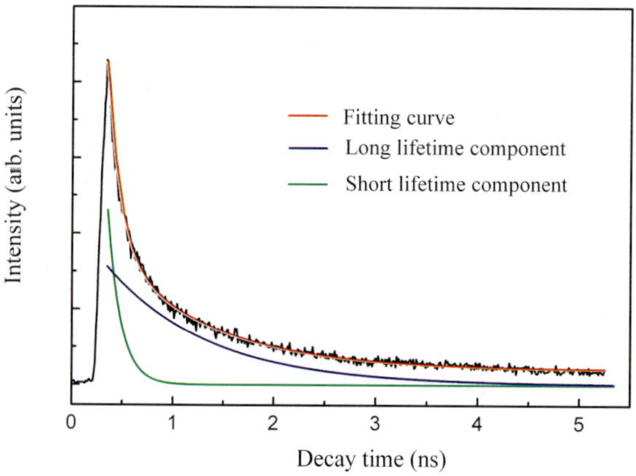

Figure 16.12 Time-resolved PL spectrum at room temperature for a Si-in-SiN_x sample. The decay process starts at 0.33 ns on the ordinate.

Such a high efficiency results from a substantial reduction of non-radiative processes in the sample, which can be confirmed by decay time measurements. Fig. 16.12 illustrates a time-resolved PL spectrum measured in a band of 140 nm around the peak position[10,13]. The decay process can be well fitted with a double-exponential function

$I \propto 1.58 * \exp(-t/0.13185) + 1.0 * \exp(-t/1.0027)$, where the time is given in nanoseconds. The two decay times of the curve, one ~ 1.0 ns and the other ~ 0.13 ns, imply the presence of excitons with two different lifetimes, responsible for the PL. Such a result represents a remarkable step forward in obtaining fast decaying PL in silicon nanostructures compared to previous reports where the decay times were generally a few nanoseconds or even microseconds, and were measured at low temperatures[29]. This heralds the possibility of fabricating silicon-based optical interconnects to operate at gigabit per second data rates. The nanosecond decay time for PL achieved here can be attributed to the fast recombination of excitons with short radii within a nanoparticle that has better passivation[30]. If QCE is the mechanism responsible for light emission then a particle size as small as 1.8 nm clearly sets a severe limit on the radius of localized excitons and consequently a faster recombination dynamic results due to the greater overlap of carrier wave functions and the direct-gap behavior.

One may tend to attribute the high-efficiency blue-violet PL and the nanosecond decay time in the current work to an impurity origin instead of the size effect in silicon particles, but this possibility can be excluded by a brief analysis. The PL centered at 428 nm was the result of our long time effort which resulted in the reduction of PL wavelength from Si-in-SiN_x films by reducing the particle size — there is an established, qualitative correlation between the blue-shifted PL and the reduced particle size[5, 6, 8-10]. Even a tiny increase of silicon particle size exhibits detectable red shift in PL, see again Fig. 16.11. In fact, even two samples grown under the same conditions, but one for 7 cycles and the other for 11 cycle, show a shift of ~ 5.0 nm in PL which we believe results from the ripening of silicon particles. Moreover, in the SiN_x samples free of any distinct silicon particles, PL showing aforementioned features is simply absent. Such observations clearly are not in agreement with the impurity mechanism alternative.

The significantly enhanced PL efficiency obtained in the current Si-in-SiN_x films is thought to be due to the sharp particle/matrix boundaries which can essentially suppress the non-radiative processes. Another beneficial feature of the Si-in-SiN_x films prepared with the cyclic growth method is the improved transparency which turns out to be beneficial for light extraction.

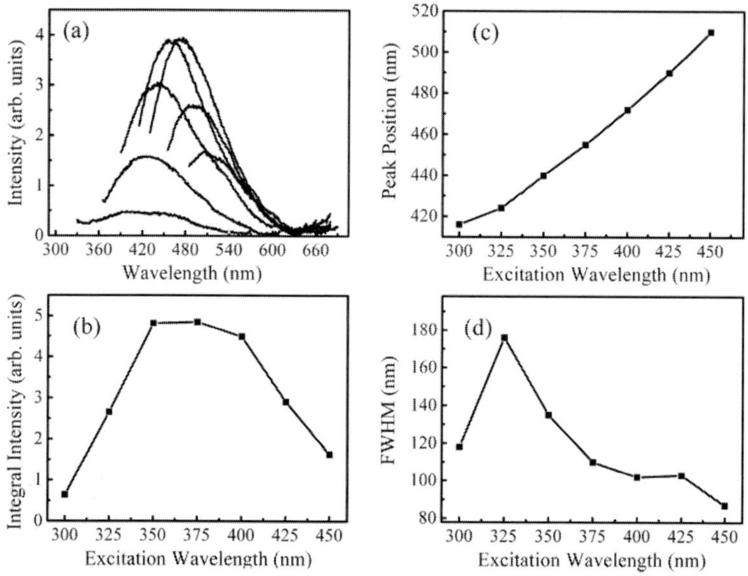

Figure 16.13 PL features as a functions of exciting photon wavelength. (a) Full PL spectral profiles; (b-d) the variation of their integral intensities, peak positions and the full-width-at-half-maximums. The excitation wavelength is set at a 25 nm interval.

Light-emission from a confined nanoparticle depends critically on such particle characteristics as size, shape, and the passivation conditions etc. For silicon particles smaller than 2.0 nm, grown by CVD method, size fluctuation is unavoidable. Based on the over-simplified model for quantum confinement where the photon energy is given by, $E = E_g + C/d^2$, where E_g is the energy gap for bulk silicon, it follows then that $\delta E \propto \delta d / d^3$. Therefore, even a small fluctuation in the d value will give rise to a much extended distribution of photon energies. Fluctuation of size and / or shape may also imply non-identical passivation at the boundary. It has an effect on various features of PL besides photon energy and, therefore, it deserves a thorough investigation. Here, a sample emitting at ~ 460 nm (2.7 eV) was examined, for example. In contrast to the situation with larger nanoparticles, here both the emission spectrum and the excitation spectrum are considerably broadened. Fig. 16.13 summarizes the detailed PL features measured with excitations at different wavelengths. A nearly unchanged integral intensity (with external quantum efficiency at ~3.0%) can be obtained by excitations ranging from 340 nm to 410 nm. As to the peak position of PL, it shifts roughly linearly from 416 nm to 510 nm, following the excitation (fig. 16.13c). The full-width-at-half-maximum (FWHM) of the PL spectral profiles, ignoring the asymmetry, shows a clear tendency to increase rapidly at larger photon energies except for the point at 300 nm (negligibly weak, hence unreliable). The extended distribution of excited states arising from size fluctuation facilitates the attainment of PL of comparable intensities in a broaden range of wavelengths (peaked from 416 to 510 nm), and under a wide spectrum of excitations (340-410 nm). This implies an enormous flexibility in choosing the excitation source and the emitted wavelengths for various applications.

16.4.3 Si-in-SiC

Up to now, high-efficiency PL across the whole of the visible light band, and established as arising from QCE, has been routinely obtained in the Si-in-SiO_x and Si-in-SiN_x composite films. However, the size limit upon the silicon nanoparticles directly implies a cutoff of the wavelength of available PL. A further reduction in the size of silicon particles is anticipated in the less oxidizing SiC matrix in which the formation of homonuclear bonds is unfavorable. This will allow the study of QCE at a size approaching 1.0 nm through PL measurements. Moreover, the realization of light-emission from a Si-in-SiC structure is of importance in its own right. Silicon carbide is the semiconductor material that is used for hard electronic devices to be used in high-temperature, high-power, high-frequency, and / or other harsh environments. A light emitting capability could further broaden its areas of applications. Up to now, light-emission from SiC-based structures has been mostly investigated in $Si_{1-y}C_y$/Si quantum wells and SiC nanocrystals. Measurement of strong PL from a Si-in-SiC composite film has not. been reported and the reason lies probably in the difficulty in engineering the silicon particles to a small enough size to be effective for quantum confined light emission.

By using the same procedures as described above, Si-in-SiC films were grown in which the minimum size of the silicon particles measured ~ 1.2 nm (Fig. 16.14). The corresponding nanoparticle density was found to be ~ 4×10^{12} cm^{-2}. It was also found that a further reduction in particle size, in contrast to the Si-in-SiN_x system, could be realized at the cost of particle density. It is noteworthy that only a minor increase in the silicon particle size, from ~ 1.2 nm to ~ 1.4 nm was observed even though the silane content in the precursor was raised by a factor of 16.

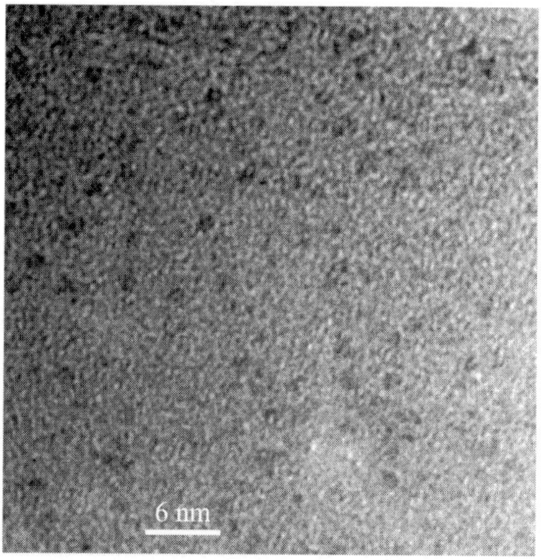

Figure 16.14 High-resolution TEM micrograph revealing the presence of innate silicon nanoparticles (black dots) in amorphous SiC matrix.

Strong PL was measured in the three samples prepared with gas ratios (in sccm) of H_2:CH_4:SiH_4=10:10:0.5, H_2:CH_4:SiH_4=10:10:2, and H_2:CH_4:SiH_4=10:5:8. With the particle size increasing from 1.2 nm to 1.4 nm, the peak position steadily shifts to longer wavelengths from 450 nm to 510 nm. The known characteristic PL emissions from defects in SiC or SiC nanoparticles were absent.

Owing to the weaker confining strength of SiC, even with the smallest particle size of 1.2 nm, the PL was found centered at 450 nm (Fig. 16.15); still a little larger than the shortest wavelength achieved with a particle size of ~ 1.8 nm in silicon nitride. Due to the extremely small size of the embedded nanoparticles, the effect of size fluctuation in defining the PL features was more severe than in the Si-in-SiN_x system. The detailed PL features under varying excitations (selected excitation between 320 nm and 490 nm) were quite similar to that shown in Fig. 16.13 (see Ref. 13). Remarkably, the emission spectrum, that extends to the violet band (below 400 nm or 3.1 eV in photon energy), loses

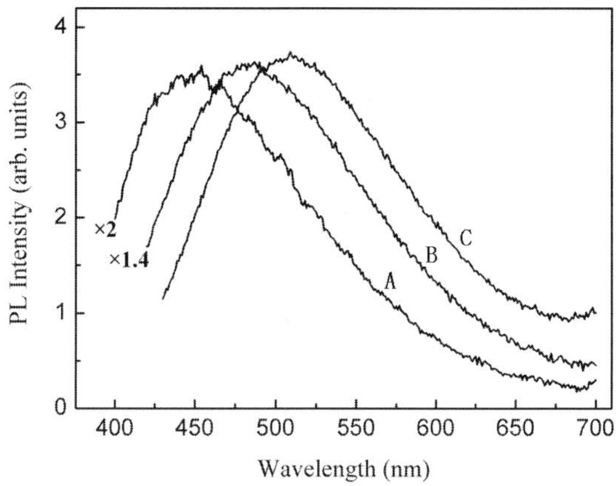

Figure 16.15 PL spectra measured for Si-in-SiC films for particle sizes in the range of 1.2 nm to 1.4 nm (from A to C). Excitation wavelength: 370 nm.

intensity rapidly, and the corresponding photon energy approached the band gap of the amorphous silicon carbide (≤ 3.3 eV). Therefore, we can say that the PL emission centered at 460 nm is of roughly the shortest wavelength attainable in the Si-in-SiC composite structure via the confinement mechanism.

Though the confined emission from Si-in-SiC system is unlikely to get down to shorter wavelengths or show higher efficiency, considering the many benefits of SiC itself as a base material for electronics, the films exhibiting strong and tunable light emission will surely find interesting applications in optoelectronic devices. For example, the Si-in-SiC composite film is useful as a working medium for UV-detection or imaging since it acts as a wear-resistant window as well. Given the broad PL emission from Si-in-SiC and Si-in-SiN_x films containing extremely small particles and the differences in PL spectral profiles from silicon particles of similar sizes, a multilayer of Si-in-SiC / Si-in-SiN_x should emit in a significantly broad band which is desirable for white-light emission. Fig. 16.16 displays the preliminary PL measurement on a Si-in-SiC / Si-in-SiN_x multilayer structure, showing extensive emissions in the green to the violet band which can be effectively excited with different wavelengths. Some interesting phenomena concerned with the electrical transport properties under light irradiation in this quantum well-dot system are now under investigation.

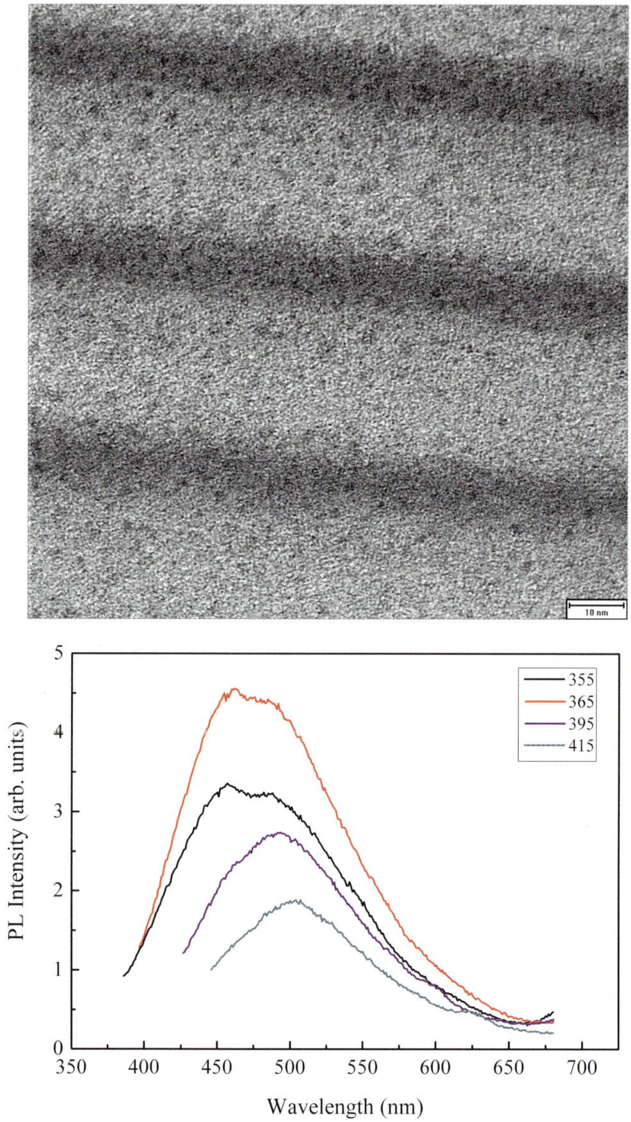

Figure 16.16 (Upper) TEM image of SiC / SiN_x multilyer structure containing silicon nanoparticles. (Lower) PL profiles of Si-in-SiC_x films measured at an excitation wavelength of (from left to right) 355, 365, 395 and 415 nm.

16.5 Conclusions

Light-emitting composite films comprising high-density innate silicon nanoparticles embedded in a silicon compound matrix including the sub-stoichiometric oxide and nitride, as well as the stoichiometric carbide, have been successfully grown. The growth was performed on cold substrates by conventional PECVD method. These are generally referred to as the Si-in-SiO_x, Si-in-SiN_x and Si-in-SiC composite structures. The silicon particles are formed in the gas discharge, thus their size and density can be conveniently tailored by adjusting the discharge parameters, in particular through changing the relative proportions of the precursor gases. To prevent subsequent diffusion-related processes in the growing films on a substrate without intentional cooling, a cyclic growth mode can be used to maintain a low substrate temperature. This also helps promote light emission by improving the film transparency.

The ultimate size and density of silicon particles embedded in the resulting composite films can be effectively tailored but generally in a coupled way. The smallest particle size we have achieved with this method is ~ 2.2 nm in the oxide, ~ 1.8 nm in the nitride, and ~ 1.2 nm in the carbide matrix. Light emission across the whole of the visible light range has been observed in both the Si-in-SiO_x and Si-in-SiN_x structures. The Si-in-SiN_x system is the best candidate for an active light-emitting medium because in this system we achieved the shortest wavelength (428 nm), the highest external quantum efficiency (over 3%), and the shortest decay time (less than 1 ns). Strong PL can also be obtained in the Si-in-SiO_x system by RTA at a moderate temperature of 500°C. However, for the Si-in-SiN_x and Si-in-SiC composite structures, the strong emission has been obtained simply in the as-deposited films.

These results show the feasibility of fabricating silicon nanocrystal-based light-emitters with a procedure that is entirely compatible with modern very large scale integration technology. A composite structure containing innate silicon particles of varying sizes or embedded in different matrices, when properly tailored, can also be used for ultraviolet radiation detection and for white light displays. Furthermore, using such highly luminescent samples, an unambiguous correlation between the particle statistics and the emission spectra may be established, in order to help clarify the luminescence mechanism.

References

1. M. Wilson 2005 *Physics Today* **April** 19-20
2. G. Hodes 2007 *Advanced Materials* **19** 639-655
3. P. M. Fauchet 2005 *Mater. Today* **8(1)** 26 -33
4. L. Pavesi 2005 *Mater. Today* **8(1)** 18-25
5. Y. Q. Wang; Z. X. Cao 2003 *Nanotechnology* **14** 1235-1238
6. Y. Q .Wang; Y. G. Wang; L. Cao; Z. X. Cao 2003 *Appl. Phys. Lett.* **83** 3474-3476
7. L .B. Ma; C. Liu; Y. Q. Wang; Z. X. Cao 2004 *JVST B* **22** 2654-2657
8. C. Liu; C. R. Li; A. L. Ji; L. B. Ma; Y. Q. Wang; Z. X. Cao_2005 *Nanotechnology* **16** 940-943
9. C. Liu; C. R. Li; A. J. Ji; L. B. Ma; Y. Q. Wang; Z. X. Cao 2005 *Appl. Phys. Lett.* **86** 223111-(1-3)
10. L. B. Ma; R. Song; Y. M. Miao; C. R. Li; Y. Du; Y. Q. Wang; Z. X. Cao 2006 *Appl. Phys. Lett.* **88** 093102-(1-3)
11. Z. X. Cao; R. Song; L. B. Ma; Y. Du; A. L. Ji; Y. Q. Wang 2006 *Nanotechnology* **17** 2073-2077
12. L. F. Bian; C. G. Zhang; W. D. Chen; C. C. Hsu; L. B. Ma; R. Song; Z. X. Cao 2007 *Optical Materials* **29 (8)** 1071-1074
13. L. Ma; R. Song; R. Huang; Y. Du; J. P. Ye; Y. Lin; Y. Q. Wang 2007 *JLUM* **126(2)** 536-540
14. S. Y. Seo; K. S. Cho; J. H. Shin 2004 *Appl. Phys. Lett.* **84** 717-719
15. F. Lacona; C. Bongiorno; C. Spinella; S. Boninelli; F. Priolo 2004 *J. Appl. Phys.* **95** 3723-3732
16. M. L. Terranova; S. Piccirillo; V. Sessa; S. Botti; M. Rossi 1999 *Appl. Phys. Lett.* **74** 3146-3148
17. L. T. Canham 1990 *Appl. Phys. Lett.* **57** 1046-1048
18. M. Zacharias; J. Heitmann; R. Scholz; K. Kahler; M. Schmidt; J. Bläsing 2002 *Appl. Phys. Lett.* **80** 661-663
19. M. A. Childs; A. Gallagher 2000 *J. Appl. Phys.* **87** 1076-1085
20. M. Shiratani; K. Koga; Y. Watanabe 2003 *Thin Solid Films* **427** 1-5
21. D. L. Lockwood; Z. H. Lu; J. M. Baribeau 1996 *Phys. Rev. Lett.* **76** 539-541
22. S. S. Iyer; Y. H. Xie 1993 *Science* **260** 40-46
23. H. Watanabe; K. Haga; T. Lohner 1993 *J. Non-Cryst Solid* **164-166** 1085-1088
24. A. Puzder; A. J. Williamson; J. C. Grossmam; G. Galli 2002 *Phys. Rev. Lett.* **88** 097401-(1-4)

25. M. V. Wolkin; J.Jorne; P. M. Fauchet; G. Allan; C. Delerue 1999 *Phys. Rev. Lett.* **82** 197-200
26. N. M. Park; C. J. Choi; T. Y. Seong; S. J. Park 2001 *Phys. Rev. Lett.* **86** 1355-1357
27. J. Heitmann; F. Mueller; L. X. Yi; M. Zacharias; D. Kovalev; F. Eichhorn 2004 *Phys. Rev. B* **69** 195309-(1-7)
28. F. Koch; V. Petrovakoch; T. Muschit 1993 *J. Luminescence* **57** 271-281
29. L. V. Dao; X. M. Wen; M. T. T. Do; P. Hannaford; E. C. Cho; Y. H. Cho; Y. D. Huang 2005 *J. Appl. Phys.* **97** 013501-(1-5)
30. K. Yamaguchi; K. Mizushima; K. Sassa 2000 *Appl. Phys. Lett.* **77** 3773-3775

INDEX